Absent Environments

Theorising Environmental Law and the City

Absent Environments offers a novel transdisciplinary approach to environmental law, its principles and its mechanics, as tested in its contextual application to the urban environment. The book traces the conceptual and material absence of communication between the human and the natural, and controversially includes such an absence within a system of law and a system of geography which effectively remain closed to environmental considerations. The theoretical springboard of the book is Niklas Luhmann's theory of autopoietic closure, whose key concepts and operations are introduced, contextualised and eventually opened up to critical analysis. Indeed, in contrast to most discussions on autopoiesis, this book proposes a radically different reading of the theory, in line with critical legal, political, sociological, urban and ecological theories, while drawing from writings by Husserl and Derrida, as well as Latour, Blanchot, Haraway, Agamben and Nancy.

In terms of environmental law, the examined concepts include environmental risk, the precautionary principle, intergenerational equity, environmental rights and urban waste, as well as discourses on community, nature, science and identity. In terms of urban geography, the city is submitted to a phenomenological analysis that builds on the existing body of work on corporeal and spatial theories, and repositions the city with regard to society, utopia, language, memory and ecological collapse. Within both law and the city, a space of absence is drawn, which signals the limitations of knowledge and accommodates thresholds of ignorance.

The book redefines the traditional foundations of environmental law and urban geography and suggests a radical way of dealing with scientific ignorance, cultural differences and environmental degradation within the perceived need for legal delivery of certainty.

Andreas Philippopoulos-Mihalopoulos, LLB, LLM, PhD is a Reader in Law, University of Westminster.

Law, Science and Society

Law's role has often been understood as one of implementing political decisions concerning the relationship between science and society. Increasingly, however, as our understanding of the complex dynamic between law, science and society deepens, this instrumental characterisation is seen to be inadequate, but as yet we have only a limited conception of what might take its place. If progress is to be made in our legal and scientific understanding of the problems society faces, then there needs to be space for innovative and radical thinking about law and science. *Law, Science and Society* is intended to provide that space.

The overarching aim of the series is to support the publication of new and ground-breaking empirical or theoretical contributions that will advance understanding between the disciplines of law, and the social, pure and applied sciences. General topics relevant to the series include studies of:

- law and the international trade in science and technology;
- risk and the regulation of science and technology;
- law, science and the environment;
- the reception of scientific discourses by law and the legal process;
- law, chaos and complexity;
- law and the brain.

General editors

John Paterson
University of Aberdeen, UK

Julian Webb
University of Warwick, UK

International advisory board

Gary Edmond, *University of New South Wales, Australia*
Timothy Earle, *Western Washington University, USA*
Fiona Haines, *University of Melbourne, Australia*
Sven-Ove Hansson, *Royal Institute of Technology, Sweden*
Sheila Jasanoff, *Harvard University, USA*
Robert Lee, *Cardiff University, UK*
Bronwen Morgan, *University of Bristol, UK*
Colin Scott, *London School of Economics, UK*
Susan Silbey, *Massachusetts Institute of Technology, USA*
Ellen Vos, *University of Maastricht, the Netherlands*

Absent Environments

Theorising Environmental Law and
the City

Andreas Philippopoulos-Mihalopoulos

R Routledge·Cavendish
Taylor & Francis Group

First published 2007
by Routledge-Cavendish
2 Park Square, Milton Park, Abingdon, Oxon OX14 4RN

Simultaneously published in the USA and Canada
by Routledge-Cavendish
270 Madison Ave, New York, NY 10016

A GlassHouse book
Routledge-Cavendish is an imprint of the Taylor & Francis Group, an informa business

Transferred to Digital Printing 2009

© 2007 Andreas Philippopoulos-Mihalopoulos

Typeset in Times by
RefineCatch Limited, Bungay, Suffolk

British Library Cataloguing in Publication Data
A catalogue record for this book is available from the British Library

Library of Congress Cataloging-in-Publication Data
Philippopoulos-Mihalopoulos, Andreas
  Absent environments : theorising environmental law and the city /
  Andreas Philippopoulos-Mihalopoulos.
    p. cm.
  Includes bibliographical references
  ISBN 978-1-84472-154-2 (hardback)
  1. Environmental law—Methodology. 2. Autopoiesis. 3. Sociology,
  Urban. 4. Urban ecology. I. Title.
  K3585.P485 2007
  344.04'6—dc22   20 06034338

ISBN10: 1-84472-154-X (hbk)
ISBN10: 0-415-57443-9 (pbk)
ISBN10: 0-203-94530-1 (ebk)

ISBN13: 978-1-84472-154-2 (hbk)
ISBN13: 978-0-415-57443-3 (pbk)
ISBN13: 978-0-203-94530-8 (ebk)

to my family
– the space in which I can be absent

In the night, everything has disappeared. This is the first night.
Here absence approaches – silence, repose, night . . . But when everything
has disappeared in the night, 'everything has disappeared' appears.
This is the *other* night.
Night is this apparition: 'everything has disappeared.'

Blanchot, 1982:163

So, why does this city exist?
What is the line that separates the inside from the outside,
the rumbling of the wheels from the scream of the wolves?

Calvino, 1993:33

C'est une ville de nulle part, pensais-je, et hors du temps.

Henry, 1977:8

Contents

Acknowledgements

Every stage of this text has involved people's presences, absences, comments and silences.

When this text was still a different text: Patrick McAuslan, Costas Douzinas, Tim Murphy, Emilios Christodoulidis, Peter Goodrich, Beverley Brown, Michael King, Louis Wolcher, Ronnie Lippens, Lisa Webley, Antonis Philippopoulos-Mihalopoulos, Marika Psaltou, Sakis my-better-third, and Andrew McGowan: in different ways, they all helped this text extend beyond itself.

When this text was being written in the garden overlooking the sea: my mother and my sister for keeping me sane.

When this text was being read and enthusiastically dissected in Copenhagen: Niels Åkerstrøm Andersen, Inger-Johanne Sand, Camilla Sløk, Anders la Cour, Bent Meier Sørensen, Birgit Vibeke Lindberg and the whole politics-management-philosophy department at the Copenhagen Business School for luring me out of my solipsism and showing me a home away from home.

When this text couldn't wait any longer: John Paterson and Julian Webb, for their confidence; Julia Chryssostalis, for her constructive doubts; the School of Law at the University of Westminster for allowing me the hours to think; and Ilias Avramidis for his loving engagement with the text and his indexing dreams.

An earlier version of Chapter 6 appeared in Ronnie Lippens (ed.), Imaginary Boundaries of Justice: Social and Legal Justice across Disciplines, Onati International Series in Law and Society, no. 11, 2005 (Hart Publishing), and we are grateful to Onati and to Hart Publishing for permission to reproduce it here.

Series editors' preface

The first book to appear in the *Law, Science and Society* series more than fulfils our early ambitions. In *Absent Environments*, Andreas Philippopoulos-Mihalopoulos presents a deep analysis of the issues surrounding environmental law, engaging with the question of what environmental law actually is and not shying away from a searching critique and radical suggestions. Inspired by, but moving beyond, Luhmann's approach to systems theory, he presents an autopoietic reading of the theory in the context of law, ecological thinking and urban geography.

Drawing as it does on earlier scientific work on complex systems, and explicitly on Maturana and Varela's concept of autopoiesis within theoretical biology, Luhmann's autopoiesis theory has obvious resonances for a series exploring the relationships between social, legal and scientific epistemologies or 'systems'. Autopoiesis offers a paradoxical model of society emerging from a circularity of operations that are geared towards the self-organisation of sub-systems such as law, economics, politics, and so on. Luhmann's work thus both challenges the post-structuralist turn in social theory and yet also breaks with classical (Parsonian) systems theory.

An increasing number of social and legal scholars are taking Luhmann's writing as a theoretical source for their work, but there are still relatively few attempts to apply Luhmann to an area of legal operations in a sustained but critical fashion. This book is a sophisticated application of the theory that both celebrates exuberantly the transdisciplinary freedom its abstraction allows and acknowledges modestly the limitations imposed by the inevitable selections and consequent omissions.

But if this is such a sophisticated autopoietic approach to environmental law and the city – insightfully recognising as it does the tendency in writing based on Luhmann's theory to make selections which enable it to support a particular position – does this mean that the current work is no more than a new description of the domain, beautiful perhaps, but of limited practical value?

If this were so, then the author would certainly be subject to censure for allowing himself the self-indulgence of producing something of purely –

though undoubted – aesthetic value in a field where we are now reminded daily that we may be sleepwalking towards disaster.

And the risk of such a situation is surely high: autopoiesis is not a ready tool for critical studies, political activism or even more mundane regulatory reform. Luhmann is famously pessimistic – though by no means depressed – about our ability to achieve the goals we set, while acknowledging our inability to let the future come as it will.

Conscious of these concerns, Philippopoulos-Mihalopoulos proposes a highly innovative and sophisticated approach, which makes this text a significant contribution to autopoiesis theory as well as environmental law. He suggests a rearrangement of basic autopoietic distinctions – albeit one that he contends can be accommodated by the theory in light of the particularity of the present field of study. One thus remains within the realm of description that Luhmann allows, but with the possibility in view that this may 'surprise' a system's self-description, allowing it perhaps to 'operationalise' the new description. The claim remains modest: the author highlights the difference between this approach and the 'bravado' of conventional social reform.

What the autopoietic approach allows, however, is a sophisticated and challenging consideration of the way in which law deals with the environment. Insofar as environmental law – for example, in the form of the precautionary principle – confronts the system with the knowledge of ignorance and the limits on action, the autopoietic approach reveals how the system responds by expelling the environment and yet simultaneously internalising this very absence 'as the embodiment of incommunicability'. This may be uncomfortable, but it may nevertheless be an accurate description of the legal system's response to ecological challenges.

This is, then, a sustained meditation on some of the most immediately apparent of society's contemporary problems, which are seen to stretch law and science. It is both challenging and original, by turns philosophical and poetic, and an excellent opening to what we hope will be an innovative and exciting series.

John Paterson
Julian Webb
January 2007

Closing

I

Although the best way to understand what this book is about would be to close it, please don't do it yet. Admittedly, 'closing' is not the most accommodating of openings. But there is no way around it. This book is all about closing in the beginning. At the end there is opening, but then again, only after closing. Throughout its course, the text performs a movement that ignores the need for beginning and end, for origin and purpose, levels of stability, clusters of directionality, mirages of advancement and chains of causality. This is an ecological book devoid of 'home' and 'logos'; an urban text devoid of narrator; and a legal text devoid of judgment. But these withdrawals can only hope to be convincing if they take as their starting point a rather peculiar opening: that of autopoietic closure.

This book is an elaboration of Niklas Luhmann's autopoietic theory as applied to law, ecological thinking, and urban geography. While firmly rooted in autopoietic thinking, it is not simply an application of Luhmann's autopoiesis to environmental law and the city, but an autopoietic interpretation of the theory. This means that my reading is an exploration of alternative strains within autopoiesis, an immanent critique as it were of the theory, which attempts to bring out different viewpoints that further describe autopoietic operations. The reason for this is partly the specific subject matter, namely the ecological considerations of the protection of the environment as seen through environmental law and the city; and partly, my wish to demonstrate the relevance of autopoiesis as the theory that captures transdisciplinary uncertainty and contingency more effectively than other current theoretical conceptualisations. The peculiarities of both my subject matter and my theoretical desire, however, call for a reading of autopoiesis that regularly departs from Luhmannian autopoiesis and in so doing, engenders further spaces of doubt within the theory.

The theory of autopoiesis is a rewardingly complex construction, whose foundational concepts and operations are explained in this book. As a taste of what is to follow, the reading of autopoiesis adopted here is one that describes (autopoiesis describing) the world in the form of a continuous rupture. The theory takes as its starting point the closure resulting from an arbitrary rupture between observer and observed. This rupture is continuous: not only does it repeat itself relentlessly and uniquely (every time as another 'once'), but in its very continuum, it allows for a bleeding between observer and observed that, paradoxically, goes against the rupture itself. In that sense, rupture and continuum are mutually undercutting: the rupture is continuous, inasmuch as the continuum is ruptured. What remains is neither closure nor openness, but the movement between the two, a bleeding into each other that translates into a trauma, visibly absent while clumsily hiding behind its mark. For, indeed, all this crossing takes place 'impassionately': autopoiesis does not have room for the use of 'good' or 'bad' epithets. The theory of autopoiesis offers no solutions, only descriptions. An autopoet uses the autopoietic tools to work through the descriptions and provide, again no solutions but more descriptions. Indeed, autopoiesis promises neither heaven nor hell as a solution; as a description, autopoiesis is a polytheistic theology whose deities construct themselves, and its proselytes are to question not only the deities, but also the whole process of questioning. Upon entering, the visitor will have to shed much of her panoply and resemiologise most of the usual weapons such as identity, origin, teleology, causality, time, evolution, memory, cognition and so on.

II

The focus of the book is double: environmental law and the city. In the first instance, an autopoietic take on environmental law does not present great difficulties since Luhmann has worked on legal autopoiesis at several instances and has often combined law with ecological considerations. Difficulties emerge only after environmental law has been described in maverick terms which do not quite fit in a Luhmannian autopoietic description, as I show in the first chapter. Perhaps the main consequence of such a description, however, is the emerged need to contextualise the operations of environmental law. For this purpose, I am suggesting the city in its abstraction.[1] The choice is arbitrary to the extent that I could have chosen an

---

1 However abstract, the city of the book is a western-type urban formation. A discussion on the
   city that would deal equally with cities in the developed and less developed parts of the

environmental NGO, the earth, or the back garden. It is not arbitrary to the extent that the city finds itself in a peculiar relation with its environment, analogous to the one through which environmental law describes itself. It is also a locus of intensification of environmental problems and legal applications. Finally, it can operate as a canvas on which various autopoietic operations can be studied and stretched, thereby instituting itself as a micro-society that contains everything, in presence or absence. This means that both environmental law and the city construct their identity (more precisely: difference) through a relation of simultaneous continuum and rupture with their environment.

The word *environment* is an abstraction. In autopoiesis, each object of analysis is a system, and a system is defined as the difference between itself and its environment. In that sense, environment is the other side of identity – the difference of identity. It is nuanced contingently and in oppositional relation to the system whose environment it is. But at the same time, the environment for environmental law is the natural environment, an object of protection, through which environmental law is defined antithetically. Finally, the environment of the city is neither inside nor outside the city, but a stage of difference between the human and the natural. By putting these three significations together, I dress the natural environment with the ambiguity, contingency and fluidity of an autopoietic proper environment and ultimately merge them into the same kind of operational treatment.

The one thing that links the various significations of the environment is absence. The environment (of the system, of environmental law, of the city) is defined in its absence (from the system), and in the subsequent way in which the system conceptualises absence. Absence is seen as the unmappable, inexplorable, non-domesticated space of ignorance within the system, the very limits and limitations of knowledge, but as encountered within the system, invited by the system and understood only as the thing that cannot be understood.

III

At first impression, especially in view of the liquid descriptions of both environmental law and the city, autopoiesis is an odd theoretical choice. There is little doubt that the theory appears obstreperous and unconventional, and that the complex terminology, as I show in Chapter 1, obfuscates

---

world would not have been possible, due to the different take on matter and meaning the latter cities would call for, which for environmental purposes can only be defined in contradistinction to the western view because of the way international economic and technological balances are established. See Neves, 2001.

application. But there is also little doubt that this theory can become aston-ishingly personal. A reader who has managed to go through the semantic jungle of autopoietic terminology may encounter a moment of 'transcend-ence', a crossing of sorts not even envisaged by the theory itself.[2] The more one proceeds into the theory, the more one catches oneself understanding the process of understanding autopoietically. Autopoiesis itself is not left out of this either: autopoiesis can only be understood autopoietically. A certain overlap between epistemology and ontology, combined with high levels of abstraction, makes the connection between autopoiesis and Husserlian phe-nomenology more than simply passing – a connection pointed out through-out the text when appropriate.[3]

Still, amidst what may sound an unholy addiction to some chemically – or biologically! – induced state of mind, I can muster the required lucidity to enumerate the three pragmatic reasons for which I chose to approach environmental law and the city autopoietically: first, autopoiesis offers as a starting kit the needed illusion of 'stability' for a relatively new and untested legal discipline still trying to find its place within the broader legal system, and especially in its particular contextualisation in the urban topology. The same applies to the city, which, although established and crystallised, is presently encountering new ecological risks and threats which ordinarily escape conceptualisation and become invisibilised within an overwhelmingly constructed environment. In this respect, autopoiesis offers the tools for a conceptualisation of ecological issues that markedly differs from the traditional anthropocentrism versus ecocentrism discussions. In so doing, autopoiesis manages, better than other contemporary theories (but within its limits), to describe the structures of ignorance within society.

Second, autopoiesis primarily *describes* the world, and only subtly and modestly, rather than ambitiously,[4] does it suggest the possibility of change. This does not mean that I do not reserve a special place for utopia in the analysis. On the contrary, in some respects the whole book is a utopian analy-sis (performatively elusive, atopical, never to be fully encountered). But the

2 In fact, transcendence is a state strictly discouraged by Luhmannian autopoiesis, which takes especial pride in its non-metaphysical, non-transcendent deployment. Although one could be forgiven for forgetting it, one must bear in mind that this is a sociological theory after all.
3 The relation between autopoiesis and phenomenology is more than just peripheral; however, a comparison eschews the purposes of this work. While the general points of confluence are mentioned at appropriate instances, at present it is perhaps pertinent to point out that their methodological and epistemological similarity can be attributed to the fact that, because of their inherent complexity, they both describe the world (their initial object of description) and themselves (their proposed methodology) in equal measure, not least because the world, according to both, can only be described through a certain lens, be this the systemic observa-tion or the philosophically informed bracketing. For a comparison between autopoiesis and phenomenology from a neurological point of view, see Petitot *et al.*, 1999.
4 King and Schütz, 1994.

utopia suggested here, the autopoietic utopia, is a contained, recursive, oscillating utopia based on the awareness that, whatever the points of departure and arrival may be, they remain a description. In autopoiesis any change starts from and finishes in a description of the system and by the system, and this is arguably the greatest utopian element of this book.[5]

Finally, the third reason for which I find autopoiesis temptingly challenging originates in Luhmann's description of environmental law that can be found in his environmental-specific book *Ecological Communication*.[6] Luhmann refers to environmental law as little more than a cut into other, pre-existing legal domains, such as planning law, police law, tax law, etc. At the same time, Luhmann himself acknowledges the potential incongruence between environmental problems and legal categories, and a consequent arbitrariness in environmentally related decisions, especially when thresholds have to be defined and risks to be assumed.[7] This 'arbitrariness' is what I want to explore and contextualise for the purposes of an autopoietic reading of environmental law. What for Luhmann is a problem, for me is the answer. Through the lack of communication between society and environment, Luhmann extracts the notion of a reinforced boundary.[8] Where I differ is that I see such arbitrariness and incommunicability as potential means of 'convergence' between environmental law and natural environment, and opportunities for a redefinition of environmental law, no longer as a 'cut into' something pre-existing but as a fully-fledged instance of autopoietic evolution with peculiarities that denote its idiosyncratic operations.

In this book, Luhmann is in good company. Derrida and Husserl appear frequently, together with a host of critical theorists from areas as varied as philosophy, geography, politics, ecofeminism, post-ecologism, literature, and so on. This, however, does not make the text a comparative work since I have not systematically consulted any other theorist. Thus, Husserl is mainly employed as a demonstration of the possibility of connecting and transferring Luhmann's social theory to other fields of analysis (such as the environmental, the human, the material). This is facilitated by the fact that Luhmann has often relied on Husserlian phenomenology, although the connection has been systematically disregarded. Derrida, on the other hand, along with other post-structuralist texts, is used as a mirror against which some of Luhmann's texts are erected. The juxtaposition reinforces the self-questioning of autopoietic theory and, to my delight, expands its already

---

5 Every development is, in Luhmann's words, 1989b:67, 'bound to corresponding points of departure'.

6 Luhmann, 1989b.

7 A point also taken by some commentators: see, e.g., Munch, 1992, who points to the problems created by environmental complexities that elude the differentiation between legal right and wrong.

8 Luhmann, 1989b:80.

considerable spaces of uncertainty. The links between autopoiesis and post-structural thinking appear in various degrees of visibility throughout the text: in a performative fashion, the text takes the reader deeper in a reading of autopoiesis which fragments itself and embraces its paradoxes in a way that is more at home in deconstruction than systems theory. Even so, autopoiesis is read autopoietically, with the result that uncertainty is emphasised and employed exponentially throughout the text. In short, my loyalty lies neither with Luhmann nor without. My interest is to employ autopoiesis as arguably the most representative theory of uncertainty that exists today (and I am fully conscious of the magnitude of such an observation), and to disrupt it further so it can accommodate a different conceptualisation of uncertainty, more amenable to the needs of such systems that put into doubt even their own systemic nature. In other words, what I try to do is offer a description of autopoiesis that undermines itself and its very foundations, thus revealing aspects of the theory that could be employed and deployed in contexts not initially envisaged. Of course, autopoietic theory, especially in the legal area, has been supremely advanced towards a plural explosion, both substantial and epistemological. Thus, Gunther Teubner's work, apart from being one of the main entry points to autopoiesis for the English-speaking audience, has managed to expand Luhmannian autopoiesis both by disengaging it from a statist, positivist impression, and by siding it up with other theoretical tendencies. While Teubner's writings have been a constant inspiration for this book, and indeed references can be found throughout the text, I have decided (or at least, I tried) to follow Luhmann, and work my text through and against mainly his writings. This is because my research is not confined to law, but expands to other disciplines, on which Luhmann has written more extensively (and when he has not written, is equally telling, for example, geography or community); and second, because I felt that I needed to return to the 'origin' rather than try to consolidate my stance against an already accomplished reading of Luhmann's autopoiesis, itself escaping the confines of simple interpretation and approaching the foundations of an established school of thought.[9]

IV

The main purpose of the book is to address ecological considerations through a theory of absence. Absence in this sense equates with the ignorance reserved for the mechanics of the relation between, on the one hand, environmental law and the city, and on the other, the environment. This

---

9 Teubner, 1993, has provided an exhaustive treatise on law as an autopoietic system, in parallel to Luhmann's position, thereby creating a current of thinking quite different to Luhmann's.

conceptualisation builds on a post-ecological paradigm that accepts the need for environmental protection on the basis of absence of its object of protection. Thus, we do not know what the natural environment is, whether the urban environment can be considered natural, whether human activity is unnatural, whether the law is anthropocentric or ecocentric, where city stops and nature starts, where law stops and ethics begin, when reality meets utopia. This space of encounter between various strands of ignorance is what I conceptualise as absence. Absence is ignorance, and must remain so, if anything is to be done about it.

By twisting and imploding the theory of autopoiesis, I show how the environment can be examined from the perspective of absence. This absence is included in autopoietic descriptions in a manner neither colonising, nor all-descriptive. The connection between the system and its absent environment is gradually unfolded through the structure of the book. The latter is straight-forward: the first chapters are reconnoitring and explore autopoiesis and its ramifications for the chosen systems of observation, while the rest are more specific as to the legal and social concepts they tackle. The first chapter deals with autopoietic law. The premises of Luhmann's autopoietic theory are explained with special reference to the legal system. The first grains of 'resistance' are planted here, when I explain the peculiarities of environmental law and the potential problems of incompatibility with autopoiesis. The protection of the environment is taken to be the 'object' of environmental law. However, nature is absent from society, since there can be no communication between them. The problem, therefore, is initially sketched as follows: how can a system occupy itself with absence? What is the relation between a system and its inaccessible environment? To this effect, I suggest an elaboration of the relation between system and environment on the basis of the form continuum/rupture, where system is its environment and vice versa. This introduces a space of absent alterity within the system, which escapes the usual Luhmannian mechanisms of external reference, and proceeds in the limitation of the system within and by itself.

This return to the system, which augurs a more solipsistic focus than that of Luhmann's, calls for a rethinking of the concept of society in its role as a supra-system. This is dealt with in the second chapter, which sets out to describe the city as an autopoietic system. The city is a departure from various aspects of the autopoietic dogma, establishing a boundary selection between social and conscious, and describing itself as a chain of moments of intensification that are built around absence. In this process, issues of reality and utopia are addressed, as well as the role of space and body in a potential autopoietic description of the city. The chapter as a whole elaborates on the form continuum/rupture, by locating the limits of ignorance firmly within the system and its illusions of knowledge.

In the third chapter, the two systems of environmental law and the city are brought together through a process of reciprocal invitation to conditioning.

This is not a straightforward autopoietic process of coupling of structures, but one that has shifted its focus from the system to the absence of the environment. In other words, it is the very absence of the environment that calls for a coupling with the other absence, thus creating a reciprocity of ignorance which reinforces the understanding of systemic limitations. This commonality of absence is tested through a negative use of silence and lack of communication, as well as an acceptance of the paradox as the form (continuum/rupture, system/environment, presence/absence) that poses the question whether there is an ethical call within the system.

The negation is further reinforced in the following three chapters that deal with more specific applications of environmental law in the urban context. In a way, the formulation of these chapters follows a temporal pattern. Thus, the first of these chapters (Chapter 4) introduces the temporal dimension of systems, and explores the future through environmental risk and the precautionary principle – a fundamental principle of environmental law that deals specifically with scientific uncertainty and ecological ignorance. Risk is defined autopoietically as originating and existing in the system (the city, the law), and as such is conceptualised as the space of ignorance within. The system deals with its future risk by projecting its present onto this space, and then withdrawing from it, in recognition of its limitations. Here, issues of science, politics and subjectivism are discussed and contextualised in the general discussion on risk prevention.

Absence is contextualised differently in the fifth chapter, where community, as the withdrawal from space, clusters itself around its very absence. An ecological community is defined on the basis of absence (of commonality, of environment, of future generations), and is connected to a rights discourse that accommodates environmental considerations. This chapter constitutes the greatest departure from Luhmannian autopoiesis (Luhmann did not have much to say about community), and into a construction of autopoietic theory based on community writings by Nancy, Blanchot and Agamben. The outcome is an even more than usual self-undercutting autopoiesis, which balances between a withdrawal from nostalgia and a loss of singularity.

The final chapter reaches to the past and shows how the latter is autopoietically connected with the present and future through the concepts of memory and cognition. Here, the environment is seen both as the space of deposit of urban waste and as the space of forgetting in the autopoietic process of being-becoming. While I tackle the practicalities of urban waste from the legal point of view, I attempt a link between the distanced waste and the non-selected part of legal decisions, and I argue for a memory of forgetting as a way of making absence visible (but always as absence) to the system.

Chapter 1

The law

Autopoiesis and environmental law

This chapter starts rather ambitiously. It attempts to describe briefly the autopoietic theory of Niklas Luhmann. Its only consolation is that it does not purport to be either exhaustive, or the only part of the book that attempts the above. Indeed, aspects of the theory are discussed as they come up throughout the text. The purpose of this chapter, therefore, is to offer some preliminary remarks on the basic notions that Luhmann uses in his autopoietic discussions – thereby concentrating more on the post-1980s writings where autopoiesis features as an integrated paradigm in the theory.[1] Because of their abstraction and generality, these notions tend to inform several areas on which Luhmann has written, including law. Thus, the following discussion is structured bearing in mind the potential relevance of the notions both to the legal system and to what follows in the rest of the book.

The exposition of Luhmann's autopoiesis finds a more concrete expression in the following section, when contextualised in the sui generis *nature of environmental law. Environmental law is presented in rather maverick terms, as a discipline that relativises autopoietic descriptions. Such an observation, however, does not lead away from closure; on the contrary, I delve deeper into it by attempting to reconcile the two. The idiosyncrasies of environmental law are interpreted from an autopoietic view, and with this, the first seeds of the present theory are planted: how can an autopoietic system internalise its environment in its absence? The question is particularly relevant to environmental law because of its peculiar connection with the environment – both systemic and as an object of protection. An inclusion in asymmetry is suggested here as internalised absence, which operates as the first manifestation of the form continuum/rupture that runs through the whole argument of the book.*

I AUTOPOIESIS AND THE LAW

Niklas Luhmann's autopoietic theory has been characterised as 'a moving target',[2] not only because of its numerous mutations and its spiralling variety

---

1 Luhmann shifted from using Parsonian methodology to autopoiesis after 1982; see Luhmann, 1982b.
2 Jacobson, n.56, 1989:1664.

of pronouncements, which depend on the field to which they are meant to apply,[3] but also because of Luhmann's ingenious way of progressing from idea to idea, assuming and incorporating change without ever being incoherent – although consistently being purposefully paradoxical. This quantum quality of Luhmann's writing is only one of the problems one faces when dealing with the theory and certainly when trying to present even its most rudimentary premises in a few pages. Indeed, if Luhmann's theory is justly characterised 'ambitiously modest',[4] then my attempt to present it here must be absurdly ambitious if not futile. Still, this chapter should be read as an introduction to Luhmann's autopoietic concepts, which will be enriched with new concepts and more elaboration in the course of the book.

Luhmann's legal theory of autopoiesis is founded on the reverential absorption of masters such as Weber, Marx, Durkheim, Freud and Parsons, and on the traditional legal currents of naturalism and positivism, enveloped in an occasionally baffling blanket of biological inspiration. Interestingly, the above theories are so well integrated and seemingly a-politicised within Luhmann's oeuvre that they appear to have lost their origin – or that their origin has lost its usual importance – and become cogs of a great machine or cells in the service of the indefatigable autopoietic organism. In this respect, Luhmann's writing is performative, painstakingly describing its descriptions.[5] Indeed, if autopoiesis is a biological theory describing the operation of the living cell, Luhmann's writing emulates this biological quality by constructing its poetics in a typically autopoietic way: the text – just as the theory – has no origin or *telos*, acknowledges no barriers to the incessant flow of ideas, epistemological tools and conceptual loans, and accepts no containment of its abstraction into gnostic formulae; nevertheless, the text – just as the theory – rarely shakes its solid sociological structure for the pleasures of transcendental poetics, despite the constant *sous rature* of its theoretical prestidigitations.[6] The most interesting aspect of this erasure, however, is that, not only the text, but also its author are constantly elusive. Just like his text, Luhmann reconstructs his thought by changing his tools without ever relinquishing his identity.[7] As put by Jacobson, 'almost anything one says

---

3 Although Luhmann's theory is predominantly sociological, Luhmann has written extensively on law, economy, politics, religion, poetry, environment, art and love amongst other things.
4 King and Schütz, 1994.
5 Luhmann, 2002a.
6 'under erasure': a manifestation of Derridean *différance*, 1976a; the relation between Luhmann and Derrida will be hinted at and, when appropriate, analysed throughout the text. The marriage is not indefensible: Cary Wolfe, 1998:144, calls autopoiesis 'the reconstruction of deconstruction'. Luhmann has repeatedly dealt with deconstruction and Derrida specifically (e.g., Luhmann, 2002f), although never vice versa, which peculiarly advances Luhmann's position as an (outstandingly informed) outsider to continental philosophy.
7 Or at least his approach to identity: Luhmann, 2002b.

about Luhmann's thought is bound to be wrong, since he, like quantum reality, leaves a position as soon as one observes him taking it. In other words, he is a great theorist'.[8]

The origins of Luhmannian autopoiesis lie in the biological determinism of Umberto Maturana and Francisco Varela.[9] The word 'autopoiesis' (from the Greek words 'auto', which means 'self', and 'poiesis', which means 'creation') has been coined by Maturana to describe the process of self-production which is the main characteristic of autopoietic systems. Autopoiesis is a fusion (and an advancement) of the teachings of systems theory with the operations of living organisms. The principal offspring of the combination between systems theory and autopoiesis is the concept of *autopoietic system*, which is a system that reproduces itself, its constituent elements, and its processes of reproduction.[10] The consequence of such a circularity is that the system recognises no *telos*; there is no inherent purpose in the system except for its own being. In autopoiesis, being is becoming.[11] The tautology between the given of existence and the beyond of creation constitute the cornerstone of the autopoietic paradox.

In autopoiesis, everything worth observing is a system. Society is a system (the 'suprasystem'). The law is a system (a 'subsystem'). Everything included in society is a system, and this everything is laid out for an observer to observe.[12] For Maturana and Varela, the observer has a clearly defined role as the epistemological authority – unquestionable within her subjectivity[13] – who brings forth the observed entity: '[a]nything said is said by an observer'.[14] Quantum-like, the observed changes when observed. But does it *really* change or is it the perception of the observer that defines (itself autopoietically) what is going to be perceived and consequently changed? Instead of giving a definitive answer, Maturana and Varela point to the *intentionality* of knowledge – thereby throwing into relief one of the commonalities between

---

8 Jacobson, 1989:1664, n.56.
9 Maturana and Varela, 1972. The legitimacy of transfer of a biological theory to the social domain has been commented by van Zandt, 1992. However, biological autopoiesis, itself already couched in intentionally abstract terms, is further abstracted and developed by Luhmann, who would not consider evolution in isolation, but always as co-evolution in the sense of structural coupling – see following chapter and Teubner, 1993:52. In any case, most of the terms used by autopoiesis were shown not to have originated in biology but in other sciences: Rottleuthner, 1988.
10 Luhmann, 1995a.
11 Maturana and Varela, 1992.
12 Almost 'everything', since, first, humans are not included in society (see below), and, second, the observer can never observe himself and the world at the same time. This is the beginning of the 'blind spot', which informs the autopoietic paradox. Luhmann, 1989b.
13 Indeed, there is a divergence between biological and Luhmannian autopoiesis on the role of the observer; see Neves, 2001. Divergences, however, exist in Luhmann's writings too; see Gumbrecht, 2001.
14 Maturana and Varela, 1972:9.

autopoiesis and Husserlian phenomenology: 'the phenomenon of knowing cannot be taken as though there were "facts" or objects out there that we grasp and store in our head. The experience of anything out there is validated in a special way by the human structure, which makes possible the "thing" that arises in the description'.[15] Indeed, 'every act of knowing brings forth a world'.[16]

Although Luhmann departs in some respects from the above,[17] he maintains and develops the Husserlian influence.[18] In his attempt to do away with the division between subject and object, Luhmann institutes observation as the link between the two: the system observes as well as being observed. In such a whirlwind of observational activity, there can be no differentiation between the observer and the observed. Although Husserl admits that reality 'in itself' does exist, that objects do have an existence independent of a subject,[19] an observer (or an *agent* in Husserl's terminology) can only perceive reality as it appears to him or her: there is no point in referring to the 'objective' reality, since nobody perceives it as such. This can only mean that the focal point has to be the observer and what they do in order to perceive things (what Husserl calls *noesis*, the act of perception of the object). In combining the existence of an object (objectivity) and the act of perception of the specific object performed by an observer (subjectivity), Husserl resemiologises the Kantian concept of *intentionality*, which now denotes the directedness of consciousness towards the world: consciousness is always consciousness *of* something. Object and subject are linked with the bridge of intentionality which guarantees that neither can operate independently of the other, yet neither is reducible to the other.

Luhmann multiplies intentionality and expands it to all directions by conceding that the observer can also be the observed. This is achieved not only through the interchangeability of observation between human beings and social systems, such as law, politics, religion and so on, but predominantly through the introduction of a *second-level observation*, which focuses on observing how others, first-order observers, observe. As Luhmann puts it, '*observing systems* in the double sense of the English -*ing* form. We ourselves may be observing systems observing observing systems'.[20] The level of second-order observation has become Luhmann's favourite locus of problem describing and solving, as well as his version of biological

---

15 Maturana and Varela, 1992:25.
16 Maturana and Varela, 1992:25.
17 Indeed, Luhmann departs significantly in many respects from Maturana. For an excellent analysis of their differentiated identity, see Rodriguez and Torres, 2003.
18 Luhmann's influence by Husserlian phenomenology is generally infrequently acknowledged in the theory, although regularly acknowledged by Luhmann, both in occasional references in the text and (mainly) footnotes, as well as in his 2002c study of Husserl's Vienna lectures.
19 Husserl, 1970a.
20 Luhmann, 1992a:1420.

autopoietic 'objectivity'.[21] At the same time, not unlike Husserl's 'higher' level of apprehension of reality, namely the famous 'transcendental attitude', Luhmannian second-order observation is suggested as the appropriate level of describing society, avoiding (while including, through the abilities of a nearly psychoanalytical observer) psychologism, and replacing metaphysics with a seemingly grounded sociological inquiry that does not accept a place for the sublime, the transcendent or the 'outside'.

This way of merging observer and observed is similar to Edgar Morin's epistemological foundations: 'The indissociability of the relation between the observer/subject and the observed/object . . . leads to the necessity of including, not excluding, the observer in the observation.'[22] Indissociability replaces dualism with the concept of observation. Hence, the concept of the observer embodies indiscriminately both social and conscious systems (as human beings are referred to in autopoiesis).[23] By merging the subject and the object under one common characteristic, Luhmann follows and develops the Husserlian revolt against Cartesian dualism, and builds the theory of autopoiesis, not on the specific traits of each system, but on the processes that conceptually 'link' while differentiating these systems.[24] By founding the theory on an abstract stratum of procedures that apply in the manner of a continuum to every system, while differentiating the system from other systems, Luhmann builds on the all-informing premises of systems theory and attempts to formulate a *passe-par-tout* mechanism of description which uses the same tools regardless of the gnostic domain.[25]

In view of the above, it is difficult to deny that Luhmann's theory falls into the grand, totalising theory category in the mode of Lyotard's descriptions of

---

21 Compare Luhmann, 1985a to Luhmann, 2004 and the pivotal position of the second-order observer in the latter.
22 Morin, 1992:379.
23 See esp. Paterson, 1996.
24 This link, however, is not a connection. See Luhmann, 1988a, where Luhmann explicitly refers to the rejection of the difference between subject and object and its replacement by the operations of self-reproduction and observation. One could be helped to think of observation as the *intentional* link between observer and observed – note, however, Luhmann's (2002c) disinterest in consciousness, which is Husserl's main starting point.
25 While systems theory has been intended to be all-informing (see Odum, 1971; Odum and Odum, 1981; von Bertalanffy, 1967, 1969; Passet, 1979), autopoiesis was initially thought by its creators to be good only for biological systems. Still, the debate whether the transposition of autopoiesis into the social domain is legitimate is devoid of any real significance for two main reasons: first, Maturana and Varela themselves have acknowledged the possibility in their later book *The Tree of Knowledge: the Biological Roots of Human Understanding*, 1992; second, while Luhmann's autopoiesis has been inspired – both linguistically and conceptually – by its biological counterpart, the two remain dramatically different in their later specialisation, mainly through the introduction of *meaning* (see below). Finally, it is interesting to note as a curio that the first one to use autopoiesis in social theory was not Luhmann but Pierro Sraffa in his *Production of Commodities by Means of Commodities*, 1960, as cited in Jacobson, 1989.

modernity,[26] or into the subjective, everything-goes category of subjectivism, where each systemic description is equally and un-hierarchically valid. While meta-theoretical second-order observation is not what I am pursuing here,[27] it is worth pointing out that Luhmann's theory is grand, but at the same time grounded not on some unifying characteristics, but on difference.[28] As I will be showing here, the very foundations of the theory undermine its grandeur: if 'grand' is to acknowledge (and include) the aporias of its own descriptions, the blind spots of observations, the black boxes of communication, the embracing of paradoxes, and the very theory's confessed inability to change the world or even a single system, then let it be grand. Every Luhmannian thesis is served together with an equally contingent antithesis, between which Luhmann's descriptions are to be found. Thesis and antithesis, however, depart from Hegelian synthesis and remain different in a ruptured continuum as actuality-potentiality, where the former is to be found (schematically) in the system, and the latter in the horizon of potentialities, all equally observant and observable, in an environment where origin, hierarchy or power is contained within systemic boundaries. This may sound eminently subjectivist, but Luhmann carefully distances himself from it and sides instead with radical constructivism, which manages to regulate the balancing act of the various systems without, however, directly influencing them.[29] But for the latter to become clear, some further observations on how the theory is structured are needed.

Systems float in opaqueness. This opaqueness is what Luhmann calls the *environment* of the system, and, as far as the system is concerned, it remains unintelligible. At this point, Luhmann departs for good from general systems theory in two ways: first, his main focus is no longer the relation *parts–whole* but the relation of a system to its environment, which is essentially a negative relation, one of differentiation rather than one of contact. The difference between system and environment is the conceptual foundation of autopoietic theory, and as such informs questions of autopoietic identity. Second, unlike general systems, autopoietic systems are not open but closed to their environment. These departures will be discussed in turn.

---

26 Lyotard, 1984.
27 It has been adequately done elsewhere, e.g., Hayles, 2000; Wolfe, 2000; Rasch, 2000; King and Thornhill, 2003; Cornell, 1992a.
28 King, 1993:223, has talked of autopoiesis as 'a total theory in the European tradition of grand theories, which extended to the whole of society and to all social systems'. The same, albeit less pronounced, is found in King and Thornhill, 2003, who talk about Luhmann's 'Sociological Enlightenment', having subsequently the difficult task of denuding the theory of the totalising theory impression.
29 The balancing act is internalised in each system in the form of functional differentiation (Luhmann, 2002d). Subjectivism is not an option since it implies that the subject actually has a choice (Luhmann, 2002b), whereas in autopoiesis, the subject (social or human) has only an illusion of a choice.

The birth of an autopoietic system is an arbitrary affair. Whether a process or an emergence,[30] it all starts with a distinction, any distinction as long as it is called 'the first distinction', the line that separates the observer from the observed, as Spencer Brown would have urged.[31] The trauma of the distinction blooms into the boundaries of a system that exists in difference to its environment. The distinction is relentlessly repeated within the system, and echoes in its environment whenever the system, in its own insular way, explores it as a means of cognitive evolution. Thus, differentiation is the beginning of the existence of any system, but also its continuation: through differentiation (from a system's environment; from other systems) a system strengthens its *selectivity*,[32] namely its ability to thematise its environment and to create more distinctions, all echoes of the initial schism. Axiomatically, however, the environment is always more complex than the system, and a system's selectivity aims to reduce this complexity – namely the infinite possibilities that the horizon of the environment represents.[33] Through selectivity, a system deals with environmental complexity and reduces it to a manageable systemic complexity, thereby differentiating itself from its environment and, hence, acquiring its identity. In autopoiesis, the schism of identity formation is exemplified in its most graphic, not only through the *sine qua non* condition of the system's *distinction* from its environment,[34] but also because of the exercise of systemic selectivity, which, for the system is an essay in *contingency*. The reduction of environmental complexity – a precondition for the system's identity – entails a selection that includes a distinction: when one chooses A over B to Z one realises the two sides of a distinction: on the one side the selected, and on the other the non-selected. Momentarily, the non-selected side of the distinction is inertisized. It returns to the scene of selections only after time has elapsed, the previous distinction has been fixed and a need for another distinction has arisen. The identity of the system is the offspring of the initial selection, for a system is nothing less than the 'interruption of continuity in the spectrum of the possible'.[35] An evolutionary anomaly? Even better: a structurally predetermined serendipity.[36]

---

30 A difference between Luhmann and Teubner, the latter supporting the view that there are stages, whereas the former radically disagreeing with it. See below, Chapter 6.
31 Spencer Brown, 1969:3.
32 Luhmann, 1982a:213. Luhmann obliges the reader by repeating and refining the general premises of autopoiesis in nearly every book he writes, making it easy to understand the theory whether approached from sociology, law, economics and so on. Indicatively, one can have a look at one of his more generic books *Social Systems*, 1995a, where all of the above can be found.
33 Luhmann, 1982a:70.
34 Cf. Fichte's, 1982, understanding of the division between self and not-self where one mutually limits the other.
35 Luhmann, 1982a:345.
36 On structural predetermination, see Maturana and Varela, 1992.

Selectivity materialises through a system-specific *binary code* which contains a positive and a negative value.[37] The legal system's code is the binarism lawful/unlawful. The legal system is functionally different to the economic, political, religious, art and so on systems, and their difference is vouchsafed by their respective binary codes. This code is only understood and applied internally, in the sense that, if the question of whether something is lawful or unlawful arises, then regardless of the outcome, this concept will belong to the legal system and as such nowhere else.[38] The binary code is the basis of the functionally differentiated nature of every system. *Functional differentiation* is an evolutionary concept, which, for Luhmann, replaces hierarchical or stratified differentiation, and manages to keep this sort of differentiation at bay.[39] The moment at which functional differentiation emerged and replaced hierarchical differentiation is Luhmann's first, Edenic distinction that pinpoints the schism between premodern and modern societies. This moment is as arbitrary as any origin, and returns to haunt modernity in the form of the threat of functional de-differentiation, with the fear of demolition of the distinct systemic boundaries and the total description of the rest of society by one system.[40]

The code is a mechanism that interrupts functional de-differentiation, and protects, say, the legal system from being directly controlled by the political, the religious or the economic system. However, the code is not a norm; it is simply a rule of causal attribution and connection.[41] Indeed, the system lacks any *Grundnorm* that would represent its unity amidst its environment.[42] Still, the binary code is a universal and invariant.[43] Everything that can be said (for the system), is said through it. Every value refers to its respective counter-value, to the exclusion of any external value. The two values of the code

---

37 Luhmann, 1989a.
38 Luhmann, 1992a.
39 Luhmann, 1995a; see also Luhmann, 1999 for a more historical description of the emergence of functional differentiation.
40 Luhmann mentions mechanisms that protect against de-differentiation. Remarkably, one of them is Human Rights – see Luhmann, 1999; and below, Chapter 5.
41 Luhmann, 1992a.
42 This is one of the foundational differences between Kelsen's *Pure Theory of Law*, 1967, and autopoiesis. The Pure Theory does have similar notions to autopoiesis, especially with regard to the nature of production of norms (a norm is produced by norms), but stops short from the conceptual sophistication of autopoiesis which presupposes neither a *Grundnorm* nor a given in the form of ultimate ground of validity, but merely a self-validating recursivity. The circularity which Kelsen attempted to avoid with the introduction of the *Grundnorm* is guiltlessly embraced by Luhmann (1985b:115) when he writes: '[d]ecisions are legally valid only on the basis of normative rules because normative rules are valid only when implemented by decisions'.
43 '[The binary code] claims universality and excludes further possibilities.' Luhmann, 1989:37. Also, Luhmann, 2004, where he points out that the unity of the system cannot be introduced in the system.

constitute a *form*, which can be defined as the before-the-distinction co-existence of the sides of a binarism.[44] A form includes both selected and non-selected space in one inoperable, pre-Edenic unity – 'inoperable' because the system cannot operate with unities, only with distinctions. But forms are deparadoxifiable and the boundaries between the values are contingently porous. The system can cross to the other side simply through selection. Thus, an operable form is a form that has materialised a selection – *either* selection between its two inclusive values. For the legal system, the code lawful/unlawful becomes operable when the system proceeds to a selection on whether something is lawful or unlawful. It is through such selections that the closure of the system is replicated. The code's positive (lawful) and negative (unlawful) value can be pictured as a switch situated on the borders of the system. The criterion is not whether the specific concept is lawful or unlawful: at this stage of 'waiting' at the system's boundaries there can be no judgment on the lawfulness or the unlawfulness of the concept. It is only after the concept has been acknowledged as part of the system's structure that the stage of characterisation arises. The binary code is the bastion of the system's autopoiesis that safeguards the distinction between system and environment (which includes all other systems), offering at the same time the flexibility of fluctuation between lawful and unlawful.

The flexibility of coding is further enhanced in two ways: firstly, the communication between the two values is allowed, in the sense that there is no obstacle in a concept being resemiologised over time as lawful when earlier had been considered unlawful.[45] Secondly, as Luhmann admits, the code is only one of a 'galaxy of distinctions' that are employed by an autopoietic system.[46] Coding is nuanced and 'connected' to exosystemic reality by what Luhmann calls *programming*, or else the procedural tool that guides the allocation of the values lawful/unlawful in specific factual situations.[47] The combination of code and programme[48] makes the legal system 'an open-ended, ongoing concern structurally requiring itself to decide how to allocate its positive or negative value'.[49] However, 'open-ended' certainly does not mean 'open'. And this is the biggest thorn in Luhmann's theory: is law really a closed system, without any contact with its environment in the usual form of input/output?

Autopoietic closure is not a simple concept. Luhmann has devoted the heftiest chapter of his *Law as a Social System* to a painstaking account of the

---

44 Luhmann, 1993; 2000.
45 Luhmann, 1989:36–43.
46 Luhmann, 2004:70, 1993b:26.
47 Luhmann, 2004 and 1998.
48 This is what Tim Murphy, 1997, 2001, calls 'framing' and 'thematisation'. The terms are interestingly endowed with fluidity and situational interchangeability, at points more so than the equivalent Luhmannian ones.
49 Luhmann, 1992:1428.

various facets of closure, and a detailed argumentation against the various criticisms.[50] It has more than often been misinterpreted as a form of extreme legal positivism or as another form of naturalism,[51] and the ensuing impossibility of interdisciplinarity (although transdisciplinarity is well accepted) is one of the contention points between critical legal studies and autopoiesis.[52] No doubt it is tempting to interpret closure as the absence of contact between the system and its environment and marvel at the absurdity of a legal system that develops without any contact with other social systems, such as economics, politics, education and so on. Such a relation of no connection has been described as 'autonomy' by critics,[53] but autopoiesis is not equivalent to autonomy.[54] To disprove such misunderstandings, two things must be clarified: first, thermodynamic considerations are not the same as autopoietic

50 Luhmann, 2004; for criticisms of closure see for example, James, 1992; Norrie, 1993; Wolfe, 1992; Lempert, 1988. For argumentation on the beneficial influence of legal closure on legitimisation, see the otherwise *contra* autopoiesis chapter on Sociological Perspectives on Legal Closure in Cotterrell, 1993. Others, such as Goodrich, 1999; Gumbrecht, 2001; Neves, 2001; Jacobson, 1989, more attuned with the conceptual sophistication of autopoiesis, argue against the plausibility, utility, empirical proof, or inherent paradox of the theory, including its concrete application to law and society in general or to genres of law, such as common law, in particular. By way of indirect but perhaps not inadvertent counter-argument to criticisms on autopoiesis and common law see Murphy, 1994b. Further, see King, 1993 and 2001, as well as King and Thornhill, 2003, for an autopoietic retort to criticisms.
51 See Cornell, 1990; Jacobson, 1989, 1992; Beck, 1994; Munch, 1992; for a supportive view see Weisberg, 1992 and Teubner, 1993. Although the earlier Luhmann was much more clearly positivist than the post-autopoiesis one (compare Luhmann, 1985a and 1993b), it would be problematic to equate autopoiesis with straightforward legal positivism: even before the introduction of autopoiesis, Luhmann's theory of closed systems never accepted the concept of the centre, or indeed the concept of a source for law (see Teubner, 1992 for one of his incisive attempts to reconcile legal pluralism with autopoiesis). By not accepting the need for beginning, by embracing the paradox and concentrating on the observation of 'how' rather than 'what', autopoiesis does away with the centralised structure of law production that positivism presupposes. As for the mechanistic underpinnings, these can easily be entertained in view of the contingency autopoiesis intentionally incorporates within its otherwise technical terminology – an uncertainty, both of effect and process, legal positivism could never accommodate without disintegrating. Naturalism, on the other hand, presupposes a pre-existing law which is discovered by a rational observer (Jacobson, 1989). However, autopoiesis neither accepts pre-existence – no *Grundnorm* – nor reserves any role for 'rationality'. What is more, autopoiesis is not structured hierarchically; there is no prioritisation among systems, norms or processes of production. The democracy of autopoiesis is a headless, nefarious and contingent evolutionary instance that can accommodate neither competition nor hierarchies among its elements.
52 Lyotard, 1984; Munch, 1992, and generally most of Vol. 13:5 of Cardozo Law Review, 1992; also Norrie, 1993; Cornell, 1990; and to some extent, Jacobson, 1989; Beck, 1994; Rottleuthner, 1989.
53 See for example Lempert, 1988.
54 Luhmann, 2004; see also Deggau, 1988; King, 2001; Luhmann, 1993a:56, where autonomy is equated to the ability for self-amendment, and Luhmann, 2004:95, where autonomy is described as a consequence of operative closure.

considerations; second, autopoietic closure is only one side of a paradoxical co-existence between closure and openness, and as such, it equates to neither stagnation nor cognitive isolation.

The issue of thermodynamics is relevant, not only because of systems theory, but mainly in view of the discussion in the following chapter, where issues of space and corporeality will be visited in the context of the urban system. It is, therefore, relevant to clarify that thermodynamics play no role in autopoiesis, be that biological or social. Already biological autopoiesis states that 'autopoietic machines do not have inputs or outputs'.[55] These belong to thermodynamic considerations, which, 'although necessarily enter in the analysis of how the components are physically constituted . . . these considerations do not enter in the characterization of the autopoietic organization'.[56] Thus, thermodynamics in autopoiesis are limited to the material basis on which the domain of observation is projected.[57] It is interesting that any surviving terminology has been carefully internalised in the system; thus, Luhmann says that systems 'use their output as input. The later development of the theory "internalized" this feedback loop and declares it to be a necessary condition of its operation',[58] as long as it remains clear that its replacement has been a consequence of the operational closure of systems. The same applies to the main unit of thermodynamics: energy.[59] The emperor's clothes have now changed: 'we can replace this concept of energy, fashionable at that time, with the concept of autopoiesis that is fashionable today', proposes Luhmann in ostrich-like irony.[60]

More concretely, *meaningful communication* is the new energy. Social systems, such as law, consist of communication.[61] Communication is the main systemic operation – thus, legal communication, political communication, scientific communication, and so on. For communication to be understood by the system, it needs to be meaningful, namely to re-establish the distinction between system and environment. In that sense, *meaning* links the system with its environment, not in the manner of energy transfer, but as 'the

---

55 Maturana and Varela, 1972:81.
56 Maturana and Varela, 1972:89.
57 Aristotle has put this distinction in the manner of substance and form. According to Aristotelian Metaphysics (1989:Chapter VII), form can be separated from substance with abstraction: the purer the concept, the less substance it contains. Form is the means by which substance becomes actual. The levels of abstraction of autopoiesis enable a system to disentangle itself conceptually from its substance (thermodynamics) and continue with its autopoiesis fuelled only by its self-production. The autopoietic self-reproduction can only be sustained if the system and its processes are seen as forms rather than substance.
58 Luhmann, 2004:79.
59 Odum, 1971; von Bertalanffy, 1969.
60 Luhmann, 2002c:54. Luhmann here refers to psychic energy, which was, however, imported in psychology from the natural sciences.
61 Luhmann, 1995a.

simultaneous presentation . . . of actuality and possibility',[62] that is as a *form* which operates in differentiated temporalities, and brings together the actuality of the system with the potentiality of the environment. When talking about meaning, Luhmann has specifically referred to Husserl's intentionality, namely the connection of difference between consciousness and the world.[63] Still, for Husserl, just as for Luhmann, there is a priority, an epistemological or ontological focus, despite sustained attempts to show otherwise: what for Husserl is consciousness, for Luhmann is the system.[64] Thus, just as the world is internally constructed by consciousness with the help of intentionality, in the same way the environment of the system is internally constructed by the system with the help of meaning.

Accordingly, meaning is system-specific: the normative rules with which a system reaches its decisions on the allocation of the positive and negative values are purely internal creations without any corresponding similarities in the system's environment.[65] The system's capacity to generate its norms that dictate its operations is the main result of operational closure. Norms generate themselves through a reflexive mechanism, namely the application of a process to itself or to a process of the same kind, very much along the lines of a DNA helix.[66] By the replication of the procedure, the system continues to create norms based on its previous norms, further creating norms on how to create norms. This last feature is the essence of the autopoietic closure. And it is in this way that a system's selectivity is heightened: selection mechanisms are themselves preselected through a mechanism of the same kind.[67]

This radical constructivist view entails, not only that the law as an autopoietic system is self-constructed, but also that the construction skills of the system extend to the environment of the system itself. Luhmann describes the environment of the legal system as framed by the system and its autopoietic operations, internally produced by the system according to its binary code.[68] As it stands, this poses problems of solipsism and factual isolation; however, the issue is solved by the intervention of a paradox – arguably the most fundamental paradox of autopoietic theory, and one that perpetually creates problems, precisely because of its counter-intuitive demands. Thus, closure relies on openness, or, to quote Edgar Morin's adage, frequently used by both

62 Luhmann, 2002e:83. Luhmann here refers expressly to Husserl's intentionality, namely the link between consciousness and the world.
63 Husserl, 1983. For an analysis of Husserl from a recondite autopoietic point of view, see Philippopoulos-Mihalopoulos, 2001.
64 The system is the only observable unit, over the environment, which, although constitutive of the system through difference of identity, it can never be observed as such.
65 Luhmann, 1995a, but through *programmes*.
66 Luhmann, 1982a:100.
67 Luhmann, 1982a:100.
68 Luhmann, 2004.

Luhmann and Teubner, 'the open rests on the closed'.[69] Closure and openness are the two sides of a form, the internal and external side of a mechanism that Luhmann calls *reference*.[70] Hence, *internal* reference (the system is what it is) and *external* reference (the system is not what it is not).[71] However, both are internally constructed by the system. Closure remains closure in the sense that there can be no input/output-type transfer of information.

Thus, through external reference, the environment of a system is introduced in the system itself, but always in accordance with the system's code. The system understands what it understands, while everything else belongs to an incomprehensible, invisible, primeval environment where no trauma of severance has ever been imposed.[72] From the environment, the system 'selects' what accords with its code, and carries on evolving cognitively while reinforcing its normative closure. This happens because every selection is a distinction between two sides of the form: external reference, namely a cognitive expedition for information which can (always already) be constructed by the system in a meaningful (for the system) way; and self-reference, namely a check for consistency of the operations of the system, always according to its already existing operations. To put it specifically in relation to law, the legal system defines itself through its legal communications, which are peculiar to the system, and through external reference in the form of negative comparison to what the system cannot identify with: politics, economy, religion, ethics, and so on – what Luhmann calls the *rejection value* that the legal system has for the codes of other systems.[73] The paradox of the form closure/openness fractally reappears in the form self/hetero-reference and the infamous '*normatively closed* but *cognitively open*' Luhmannian pronouncement.[74] Within the mechanics of the form, the paradox is thrown into relief by the fact that, no matter the selection, the two sides of the form are always mutually implied. Selection then becomes relevant in that the system is required to 'balance' the two sides of the form.[75] This balancing act is the system's autopoiesis, which can never be reduced to either closure or openness, without the understanding that the other is always already implied. The double reference has as a consequence an indefatigably contingent identity of the system. Like an organism, whose cells change but its structure remains the same, an autopoietic system constantly renews its elements without

69 '*L'ouvert s'appuie sur le fermé*', Morin, 1986:203.
70 Luhmann, 2004:86.
71 Luhmann 2004 describes even external reference as internal, since external reference is always self-reference.
72 At least by the specific system, since the environment is nothing else but a constellation of mutual lacerations between self- and external reference.
73 Luhmann, 2004:190, 1993b:187.
74 Luhmann, 1988a:20. See also Dupuy, 1988, with three suggestions as to the interpretation of openness.
75 Luhmann, 2004:97.

endangering its identity. In autopoiesis, being and becoming form an inseparable circularity: the purpose of the system is the system itself. Becoming lends
its perpetual motion to being, thereby constituting a volatile identity for
the system, changeable and unpredictable, but on a bed of solidity and
concreteness.

More specifically, the system is open to *environmental irritations* as devi-
ations from expectations the system has of the operations of other systems.
Thus, the system readjusts its operations in order to balance its contradictory
references.[76] However, and this is where closure returns, all irritations are
self-produced, in the sense that they are translated in the systemic code and
are dealt with endosystemically. At such points of irritation, changing the
systemic operations is one option; the other is to change the factual information that deviates from the system's expectation. This is not a form of
influence or control of the environment, at least not in the traditional social
engineering way, since whatever happens, happens strictly within systemic
boundaries. This is an eccentric feature of autopoiesis: a theory that appears
self-centred (if 'self' is the system, and 'centre' is anywhere to be found),
becomes immured in its own inability to thematise in any efficient way
anything else but itself[77] – and even the latter is compromised by the dual
reference of the system. Autopoiesis is characterised by, on the one hand,
an intense colonisation of the environment in the form of internally constructed reference, and on the other, an admission of the impossibility of
social control in the usual hierarchy and power-structure schema. Apart from
realistically deflating expectations, autopoiesis also shows that relations
between systems are not as unproblematised as initially thought, and that
regulation, although determining, has to be studied in an endosystemic way,
because only that regulation – the one used by the system and for the system's
perpetuation – can bring changes to the system and its environment.[78] By
delimiting this control within existing but fluctuating systemic boundaries,
autopoiesis shows how mendacious the discourse of direct social control
through law is, and indicates instead 'endosystemic', reflexive ways of social
change, based on observation, devoid of any unrealistic ambition, and with a
good chance of bringing unspectacular yet effective changes.[79]

To a great extent, legal autopoiesis is a reaction to the problems that other
legal theories have encountered so far. It attempts to mend the inability of

---

76 Luhmann, 1992a:1432. For an explanation of expectations, see below and Luhmann, 1985a.
77 Schütz, 1994.
78 King, 1993; Paterson, 2006.
79 See Teubner, 1993, for the illusions of social control through law. Also Rottleuthner,
    1988:114, who, however, approaches autopoiesis from a social-engineering point of view:
    'autopoiesis seems to be adduced particularly for explaining the use of law to give direction to
    social change; for explaining the limits of *social control by legal measures* and for specifying
    the requirements for the knowledge required for control' (original emphasis).

legal theory to answer in a satisfactory and workable way the 'big' questions, the 'metaphysics' of law: What is law? What is law becoming? Who creates law? What are its limits? Interestingly, however grand a theory autopoiesis may be, its doctrine is one of 'humility' and modesty. Instead of adding to the 'juridification of the life world',[80] autopoiesis points to the limits of law's capacities. But this has been regularly misunderstood, even when praised. Thus, often in the form of the obligatory positive comment before criticism, autopoiesis has been found to have several attractive features: it has been said for example, that legal closure is beneficial to the office of judge, legal practitioner, academic lawyer, and citizen, because it perpetuates the idea of independence with regard to the legitimacy of their decisions, actions, teachings, and expectations respectively.[81] Or, it has been suggested indeed, that law's autopoietic production (what the earlier Luhmann called *reflexivity*, which is essentially the projection of norm onto norm) is an alternative way of self-correction of the system, with tangible consequences such as increased accountability, transparency of decision-making processes, and a preference for the liberal market-based mechanisms over the traditional command and control methods.[82] However desirable some of the above may be, they are short of being representative of the intricacies of autopoiesis, not least because they focus on its raw aspects while disregarding its nuances. There is no doubt, for example, that autopoiesis propagates closure but this is only one aspect of a complex theory, and indeed one aspect of several other sociolegal theories.[83] To reside on a generalisation of one aspect of such a complex theory, however positive the comments, is unsatisfactory. For perhaps the main merit of autopoiesis is that it is a complex theory trying to describe a complex society.[84] Anything less than that would be partial. Anything more than that would be unrealistically ambitious. Hence, the theory (in theory) does not attempt solutions, only descriptions.[85]

The flipside of the above is the way autopoiesis protects its systems from themselves. In the heart of autopoiesis there is an absence of certainty, a continuous oscillation between inside/outside, self/hetero-reference, system/ environment, closure/openness. This paradoxical absence appears as a form behind every operation of the system, and remains there, invisible and

---

80 Habermas, 1985.
81 For argumentation on the beneficial influence of legal closure on legitimisation, see the otherwise *contra* autopoiesis chapter on Sociological Perspectives on Legal Closure in Cotterrell, 1995.
82 See especially Orts, 1994, 2001, for an instrumentally useful but ultimately unambitious attempt to apply autopoiesis in environmental legal practice; also, Teubner *et al.*, 1994.
83 Kelsen's Pure Theory is one example as mentioned above; also Hart's categorisation of legal rules or even the Habermasian discourse. See Cotterrell, 2001; Knodt, 1994.
84 Luhmann, 2004:67, 1993b:26.
85 Luhmann, 2004:65, 1993b:24, where he accepts that the kind of theory he proposes is not supposed to guide practice but only a description of the legal system as self-description.

unbreachable, for otherwise the system would disintegrate. In other words, the system is never really forced to face its paradox and question the very materiality of its boundaries – for what are open/closed boundaries?[86] With every selection, the form is repeated while distanced from the matter in hand. But there is a bias.[87] Even the cognitive forays of the system into its environment are more like journeys in one's own room than real adventures. External reference is the system's keyhole through which it looks at the world. Through such a reference, the world is more Disneyland than jungle. Systems are potent things, and shield themselves behind their, admittedly fluctuating, boundaries and codes. The system will take no risks if it can help it – or will it? Does the system ask itself the 'difficult' questions, or does it pass it on to other systems?[88] External reference is the domestication of the stranger, an armchair fantasy of what adventure is like. One learns from such fantasies – indeed, intelligence is largely that – but one is not traumatised by them, and it is questionable whether there can be growing up without a bit of trauma. Observation is indeed an exposure of sorts but the safety net remains. First, all observations are self-observations, since the system only 'looks' inside (which is its constructed outside) in order to develop cognitively.[89] And second, the system can never observe the environment. The environment remains the grand manqué of autopoiesis, thick with opaqueness, heavily populated with systems, but never intelligible in its own right. What is more, the autopoietic quest can never start from the environment. The system determines what the environment stands for, and while one is implied in the other, their relation is one of *englobement du contraire* ('ingestion of the contrary'),[90] an enlightened colonisation and a safe strategy of description.

II ENVIRONMENTAL LAW AS AN AUTOPOIETIC SYSTEM

Applying autopoietic theory, even on the basis of the above epigrammatic introduction to environmental law is challenging, for the paradoxical reason that environmental law lends itself to autopoietic descriptions, and simultaneously presents potentially insurmountable idiosyncrasies that cannot fit in with what an autopoietic legal system is supposed to be. Still, the choice of

---

86 A suggestion of sorts is reserved for Chapter 5, below.
87 Wagner, 1997, points to the asymmetrisation of the form by showing that Luhmann always prefers the side of the system over that of the environment.
88 Luhmann and Teubner are in disagreement: for the former, the system can decide to ignore such questions; for the latter, the decision is not so easy. See Teubner, 1993, and Smith, 2004.
89 '[t]he states of the system are exclusively determined by its own operations', Luhmann, 1992a:1424.
90 Dumont, 1966:107, cited in Luhmann, 2002c. See also Wagner, 1997.

autopoiesis as the theoretical basis for the analysis can be justified on several grounds: first, autopoiesis is an adequately complex theory to describe the complexities of environmental law; second, autopoiesis conceptualises its systems in a grounded way that, while describing potentialities, marks systemic limitations. This is particularly relevant to a system, like environmental law, whose object is in the process of definition, and as such has no in-built breaks.[91] Third, autopoiesis does not offer illusions of control and intervention, but merely observations and descriptions. This is particularly apposite to an otherwise aggressively abused branch of the legal system, whose services are variously controlled in order to fit in with more pronounced priorities, such as economics or politics. And finally, autopoietic abstraction is the most appropriate route to an insular interdisciplinarity – otherwise known as transdisciplinarity – that is needed for any adequate description of environmental law.

Thus, in the ambit of autopoietic theory, environmental law is a sub-system of the general legal system, and as such it is expected to present similar characteristics as general or otherwise specialised law. However, again in the ambit of autopoietic theory although from a potentially unorthodox point, environmental law's principal tendency is one of constant differentiation from the existing legal operations (traditions, concepts, procedures, actors, purposes and so on). There are several reasons for this double pull. First, as already mentioned, environmental law is a relatively new and as yet 'uncrystallised' branch of law.[92] Such 'infancy', however, does not mean that the law is not there. It simply means that the law has not been adequately distressed by sustained attempts at the establishment of (mutually annulling and therefore paradoxically constituting) jurisprudential underpinnings of the kind that other branches of law have.[93] On the other hand, environmental law is new but not 'fresh' – it is born with an adult skeleton, mature syntax and achieved sense of purpose. It has borrowed experience and operative readiness from other legal branches, and has proven its autopoiesis in the way it has managed to convert itself from a patchwork of communicational loans, to a self-standing legal branch. However, its foundations are shaky, not least because of the multiplicity of claiming sources of inspiration. To put it in

---

91 Environmental law has a tendency of total inclusion. This is valid for every system, but becomes more acute in the case of incipient systems without an established self-referential historicity.

92 For a theoretically informed analysis of the different functional calls for law, see Sand, 2005; for a more specific focus on GMOs that attests to the differentiated expectations with regard to environmental law, Sand, 2001; from a trade and environment perspective on the basis of systems theory, see Perez, 2004.

93 There are several texts that deal with the issue, some (e.g., Delaney, 2003; Teubner *et al.*, 1994; Gillespie, 2002) more successfully than others. Still, they do not amount to a critical mass that can confidently question itself.

autopoietic parlance, environmental law's functional differentiation is still taking place. This does not mean that environmental law cannot be seen as an autopoietic system (yet), or indeed that the present text will indulge the debate on whether there are degrees of closure.[94] While potentially relevant, the debate is not necessarily fruitful if one considers that the 'bringing forth' of an autopoietic system is an arbitrary event, a distinction which may well have some correspondence with empirical reality, but is deprived of effect in the sense of ontological intervention. The epistemological schism between system and environment is what interests me here, and how, through such an epistemology, the system can be better described in order for its actuality and, more importantly, its potentiality to be understood. Thus, when I say that environmental law's functional differentiation is still taking place, I mean that there is a scope for debate as to the foundational meaning of the system, which will be certainly more relevant and interesting than the already well-digested debates of traditional jurisprudence.

Second, environmental law is required to deal with extremely complicated matters, which instigate clashing reactions from various social domains. These reactions range from support and fanaticism to indifference and hostility. Environmental law has to play along and resist (un)fashionable environmentalism, persistent economic weighing (and equally, remuneration, since both protection of the environment and protection against the protection of the environment are lucrative), avid internalisation of environmental issues by the political system, unsettling externalisation of scientific uncertainty, ethical reflections that urge for paradigm shifts of the kind established law is unable to perform, and so on.

Third, these matters tend to make their presence felt with unprecedented urgency, imposing deadlines concerning global existence, health and quality of life. The temporality of environmental law is a hydrocephalous one, with the future weighing massively over present decisions and through institutionalised means such as responsibility to future generations. Such a distribution of juridical weight requires of the system unprecedented degrees of flexibility (or inflexibility, depending on how observant the system is of other systems' expectations of it) that pull environmental law towards domains of self-reflection untrammelled by other legal disciplines. This is multiplied by an inherent and irreducible ambiguity of environmental law, namely its exact domain of operation. Whether this is attributable to the absence of a set of principles,[95] or to the elusive nature of the issues, the point remains that it is not possible to determine with any reliable accuracy what makes a problem environmental. The other side of the same problem is an overcompensation in terms of claims of problems on behalf of the

94 Luhmann, 1988a.
95 Tarlock, 2004.

environmental legal system in order to make sure that eventualities are covered. This has the potentially useful but more often debilitating consequence of considering everything more or less environmental.

The friction between an autopoietic description of the legal system and the communication of the ecological crisis is obvious in Luhmann's *Ecological Communication*.[96] This book is, unfortunately, an unsatisfying affair. In it, Luhmann describes the general operations of several systems, and occasionally looks at how the environmental crisis is internalised by them. However, the book reads more like a consolidation (a very good one at that) of the theory at that particular stage of thinking, rather than an actual engagement with ecology. Arguably, the main issue with the book is that Luhmann keeps on vacillating between, on one hand, denying the possibility of looking at ecological problems as a problem *per se* rather than as a *resonance* within society of something that clearly is no longer of relevance;[97] and on the other, referring to a certain exteriority to society which translates in arbitrariness, insecurity, angst, and so on, without ever wondering whether this exteriority could be linked with degrees of exclusion from society, in the direction in which he has later allowed his writings to move.[98] Of course, enlightening passages occur throughout. Perhaps one of the most telling is Luhmann's admission that 'if anywhere, it is in ecological communication that society places itself in question'.[99] Luhmann concedes that, specifically in law, too much resonance of and angst about ecological issues translates into a considerable amount of unpredictability best seen when the environmental legal system needs to define thresholds and values, levels of acceptability of risk, and environmental preferences as diluted by economic and causal considerations. Of course, Luhmann continues, these are not new types of problems: there is nothing new in a system; everything is always already there. Novelties are simply another opportunity for reformulating existing operations. However, these problems 'acquire a new intensity and scope when a new ecological consciousness of the problem begins to affect the law'.[100] It is this intensity and scope that require the environmental legal system to determine its limits anew and potentially at variance with traditional legal formulations.

The above has been formulated in various ways whenever Luhmann deals with ecological considerations, most prominently in *Ecological Communication, Ecology of Ignorance and Risk*. But perhaps the most eloquent formulation is by Lindsay Farmer and Gunther Teubner: 'uncertainty about

96 Luhmann, 1989b.
97 Luhmann uses the term *resonance* in the sense of 'irritability'; 1993b:225.
98 The most representative example would be Luhmann, 1997a, but also 1995d, 1995c. But also, in his writings on religion, Luhmann, 2000b, especially as noted by Sløk, 2005.
99 Luhmann, 1989a:142.
100 Luhmann, 1989b:69.

ecological risks is irreducible'.[101] What this means is that the system is not capable of reducing the complexity of its environment for its own purposes. Still, the system observes and internalises as usual, not blown apart by the complexity that its external reference (namely, the reference to what the system is not) necessarily reports back to the system. Unless of course, the external reference does *not* really report on what is happening out there[102] – unless the external reference is simply the 'other' side of the form, the *contraire* which is submitted to the *engloblement* by the self-reference, and domesticated as it 'ought' to be for the sake of the system. A necessity that supersedes contingency? In other words, the reconstructed environment within each system – and environmental law specifically – is a palliative that the system itself constructs as such (as a palliative) in order to remain risk-free. Exposing itself (even through the mediating mechanism of external reference) to the irreducible environmental complexity of ecological risks might have entailed for the system a foundational operative shift with potentially catastrophic consequences. To put it in the environmental legal context, if environmental law were to expose itself to the irreducibility of environmental complexity, it would have to deal with a fundamental miscorrespondence between its external and self-reference (its foundational paradox), with the possible result that it would have to depart greatly from traditional characteristics of law, and venture elsewhere in search of its identity.

The *topos* of such an expedition is the main focus of the present book. For this book is not envisaging to be anything more than a locus of debate on the cognitive domain – the domain of learning expeditions – of a new and uncrystallised, yet frightfully dispersed and, at the same time, profoundly differentiated branch of law. For the time being, however, I would like to translate the issue at hand, namely the idiosyncratic nature of environmental law, into autopoietic terminology, and eventually propose a different understanding of the environment based on absent exteriority. To do this, I will look at the way Luhmann describes the function of law, and then relate it to the function of environmental law. This is best done through an analysis of the role of *expectations* in law, since law's function typically is to establish and stabilise social expectations through the handling of disappointment.[103]

---

101 Farmer and Teubner, 1994:4.
102 It is obvious that this is precisely not the role of external reference – namely a report on 'out there'. External reference is meant simply as companion to the prioritised self-reference, and as such cannot 'report' on anything.
103 Luhmann, 1985a. This thesis has later (Luhmann, 2004) been enriched by the concept of conflict as the pedestal on which law bases its stabilisation of expectations, as well as the notion of enforcement which was not considered previously. In this later version, Luhmann focuses only on normative expectations and does away with the temporal, social and material differentiation, clarifying that expectations are not individual but social. For the present purposes, I retain the older classification because it is more comprehensive, a better introduction to expectations, and a more appropriate vehicle for a discussion on environmental law.

Luhmann considers three dimensions where structures of expectations are fixed: the temporal, the social, and the material.[104] Through an analysis of these dimensions, I will address the idiosyncracies of environmental law.

According to Luhmann, reduction of environmental complexity by the system can only come in the manner of stabilisation of expectations as disappointment-proof *over time*. The temporalisation of expectations is necessary, since only thus can normativity be established (in a commonly accepted form) and transferred for future use. This presupposes – and expects from the legal system – an ability to proffer certainty of expectations which can inform social operations in the future (what Luhmann calls *time binding*).[105] Certainty presupposes the ability to construct stable boundaries between system and environment, hence the system's ability to maintain the confidence in the attribution of validity to any one term of the binarism lawful/unlawful at any one time. In other words, certainty is the ability of the system to select one of the terms of the binarism and stick to it for as long as the same or a similar question arises. This does not mean that the law can promise a conduct that conforms with norms, but simply that the law protects those who expect such conduct.[106] This kind of certainty, normally expected of the legal system, cannot be reliably provided by environmental law in view of its bias on presumptions rather than time-binding decisions.[107] This is largely attributed to the intensive and often blind internalisation of scientific findings, which are the epitome of autopoietically produced complexity.[108] The establishment of thresholds, for example, is only a minor theme compared to the formidable task of risk prevention in view of lack of exhaustive scientific information. The inherent necessity of environmental law to couple with science demands ephemeral decisions and constant re-evaluation of already established problem-solving methodologies, especially in view of the pronounced demand, within environmental law, for what appears to be an ethical binding of the future – namely, the concept of intergenerational equity.[109] 'Environmental problems are characterized by the need to reduce their inevitable uncertainty through the constant generation and application of new knowledge. They often do not, as do many other areas of the law, display a repetition of similar fact patterns.'[110] It is not only

---

104 Luhmann, 1985a:73.
105 Luhmann, 2004:146.
106 Luhmann, 2004:150.
107 'The best we can hope for are presumptions because, in the end, environmental law is a series of hypotheses that must be tested (and often modified) over a long time horizon by rigorous monitoring and experimentation' (Tarlock, 2004:220). No doubt presumptions are also time-binding decisions (and vice versa), but especially in environmental law, their ability to bind is explicitly compromised.
108 Flournoy, 1994.
109 World Commission on Environment and Development, 1987.
110 Tarlock, 2004:220.

the patterns that change, but also the scientifically-recommended way of evaluating such patterns. Environmental law is a showcase of Murphy's description of the future in the era of statistical positivisation: '[t]he future is reconstituted as inherently revisable statistical projections on a screen'.[111] The in-built need for revision of statistics renders shaky the connection between a present decision and a future stability: 'it makes little sense to agonise over today's decision when it is likely to require revision tomorrow'.[112] The inability of environmental law to fix expectations temporally, at least as adequately as other legal branches can, is not a systemic malfunction: it is simply an attribute of the system in view of the irreducible complexity of the environment.

The social dimension is the second dimension in which expectations are fixed. With this, Luhmann introduces the third party in the previous duel of expectations. If the third party has the same expectations as the primary parties, then we are referring to institutionalised expectations. Law institutionalises *normative* expectations, that is expectations which, when shared by third parties, provide a normativity to which the party can adhere despite disappointments.[113] Normative expectations rarely change: their normativity is expected to be able to stabilise expectations. *Cognitive* expectations, on the other hand, can change in the form of adaptation to disappointment. The two categories of expectations, in direct analogy to a system's structural closure and cognitive openness, are combined by the system in a form, one side of which always implies the other. From this distinction, rather than from inherent characteristics, the concept of legal norm emerges: 'the concept norm defines the one side of a form, which form also has another side.'[114] The legal system reduces complexity by fixing normative expectations that have the ability to maintain and perpetuate themselves, especially in situations of conflict.[115] Thus, although the selection of whether a disappointment will be handled normatively or cognitively is made by the system, specifically for the legal system the processing of expectations entails institutionalisation, and this occurs predominantly through normative expectations. A legal system with an overproduction of cognitive expectations is obviously not fulfilling its function very well. And this is precisely the issue with environmental law. Not only in view of scientific uncertainty, but also because of the overwhelming amount of social uncertainty that surrounds environmental issues, the law's balancing act is cognitive par excellence, and generates the same kind of

---

111 Murphy, 1997:161.
112 Farber, 1994:791.
113 Luhmann, 1985a.
114 Luhmann, 2004:149; see also Luhmann, 1993b:77–79, where he points to the correspondence between, on the one hand, normative expectations, normative closure and legal norms, and on the other, cognitive expectations, cognitive openness and legal facts.
115 Luhmann, 1989a.

expectations in society. The environmental legal system, arguably because of its incipient structures, is evolutionarily required to redefine constantly its boundaries in order to accommodate its own code/programme relation (in the sense of self/external reference, or legal decision/legal facts), such as public participation (whose inclusion always enjoys a high-profile mention in every environmental legal development), politics, ethics (especially with regard to issues of protection of the environment for its own sake), and economy, as exemplified in the concept of Sustainable Development, a legal concept that linguistically incorporates the handling of the conflict between economy and environment within the boundaries of the legal system.[116] These issues are thrown into relief through environmental law's peculiar concept of *soft law* which deliberately oscillates between binding and non-binding law.[117] And while it is indeed the case that all law has to deal with conflicting interests by internalising their meaning and dealing with them through its own systemic legal structures, the difference is that conflicts in environmental law are not only external – in which case expectations apply – but also internal, that is *between* expectations. The conflict between normative and cognitive expectations within the environmental legal system in combination with the difficulty of temporalisation, leads either to the transfer of selection to ad-hoc generic (rather than specialised environmental) judge-made law, which debilitates the normative side of the form, or to a redefinition of environmental law that presupposes less normativity than the average law, more cognitive flexibility, and significantly greater 'fuzziness' in decision making.

The uncertainty inherent in environmental law is significantly more complex and expansive than the usual 'guilty/non-guilty' binarism of the generic legal system, because of the difficulty in providing for a programme, in the Luhmannian sense of normative connection with legal facts, to go with these terms. Attributing values to the code is a matter of internal conflict which, however, presupposes the validity of certain normative selections. If normativity is temporally and socially elusive, programming will prove at best arbitrary. But arbitrary selections are, once again, a feature of environmental law, as Luhmann himself admits.[118] Arbitrariness can be understood as the presentation of an essentially cognitive selection as normative for purposes of reduction of complexity. Arbitrariness within environmental law is simply a matter of routinely systemic survival. The cardinal arbitrary selection is that of its 'object', which is also related to the third dimension in which expectations are fixed, the material dimension. The material dimension is the

---

116 Smith, 1994:1079; see also Richardson and Wood, 2006; for a critique of sustainable development in the ambits of integration, see Philippopoulos-Mihalopoulos, 2004b.
117 The concept originates in International Law – another discipline which, even at this stage, is failing to fix expectations normatively; see Rasch, 2004. According to Teubner, 1996, *lex mercatoria* is another form of soft law, hence his and his contributors' efforts to amplify its autopoietic ambit.
118 Luhmann, 1989a.

acknowledgement of the necessary community of the world in order for expectations to exist.[119] Luhmann accepts the need for a *fictional consensus* on which the reciprocal confirmation and limitation of expectations will be exercised. This consensus is material in the strict sense of the word: it consists of 'a commonality of events, visible action and symbols for the invisible'.[120] Such materiality is *de facto* impossible for environmental law. While all human-made law is human law, environmental law describes itself variably as both anthropocentric and ecocentric, that is, as having as its object environmental protection both for the benefit of humans and for the sake of the environment.[121] While the former reinstates humans as the obvious object of environmental law, the latter accepts the natural environment in the same role. In view of this potentially conflicting form, it is clear that environmental law 'is not an organic mutation of the common law, or more generally, the western legal tradition ... As a result, environmental law remains largely unintegrated into our legal system; thus, it is vulnerable to marginalization as support for environmentalism ebbs and flows.'[122] Its close connection with the fate of environmentalism aside, environmental law is a *sui generis* branch of legal communication: the problems appear with the translation of the form into anthropocentrism/ecocentrism, itself an improbable distinction, burdened by the artificiality of the division between human and natural, the insistence on subject/object distinction, and the use of the centre as the source of ontological thematisation.[123] In short, while environmental law has all the appropriate conditions to renegotiate the above distinctions and formulate itself in a manner that will facilitate the emergence of differently seen interfaces between the sides of the distinctions, the only thing that it is positioned to do within the wider supra-systems of law and society is a spasmodic incorporation of ignorance in the way of arbitrariness. Thus, no 'commonality of events, action or symbols' can exist between the main distinction at work, namely human/natural, and no consensus, however fictional, can be consistently assumed. Arbitrary selections apply to both sides of the described object of environmental law: what nature considers 'protection' is as impermeable as what environment humans consider worth protecting.[124] Arbitrariness, therefore, is necessary for the creation of the required consensus, but this time one cannot refer to Luhmann's fictional societal consensus but to a variation that goes beyond the mere licence to fiction.

119 Luhmann, 1985a.
120 Christodoulidis, 1998:124.
121 Alder and Wilkinson, 1999; Stone, 1974.
122 Tarlock, 2004:217.
123 I revisit these themes in the following section, when trying to define more specifically the relation of environmental law with its environment.
124 Latour 2004a, presents the same distinction as an opportunity for internalisation of the mutual constructions of democracy (human/non-human) in what he calls *pluriverse*.

Consensus is extended to a domain which has no signifier in autopoiesis: social systems are part of society; conscious systems are outside society (so they are *somewhere*) with occasional participation mechanisms;[125] even inclusion/exclusion is a feature that plays along the usual mechanisms of communication/perception.[126] Non-human entities are autopoietically excluded, left outside the outside, a sociology without *logos* limited to social resonance. Such semantic invisibilisation makes the imposition of consensus particularly problematic, since then one can *only* talk about consensus and not its always implied other side, namely *dissent*.[127] The emergence of differentiated ways of acquiring such consensus are a certain possibility within the horizon of environmental law's crystallisation, even if this means recognising such terms as 'human stewardship',[128] or indeed the systemic institutionalisation of arbitrariness in the form of presumption.[129]

Luhmann seems to believe that environmental law is only partially equipped to deal with the increased appearance of arbitrariness in environmental law's decision making, partly because of a relocation of responsibility to the political system – a frequent *Deus ex machina* for the legal system – and partly because of the inadequate speed at which learning takes place within the legal system.[130] While these considerations are pragmatic, they suffer from a distinct lack of uniqueness. They are not problems peculiar to environmental law, but appear equally to all functionally differentiated aspects of the legal system. Variations of degree do not make them especially apposite to environmental law, and certainly should not allow these problems to be seen as insurmountable limitations, at least not any more than those of any other legal discipline. However, environmental law's arbitrariness is different in that, due to the mismatch between self- and external reference it entails, it induces a constant need for consistency checks, which itself amounts to a space of ignorance within the system. Undoubtedly, the systemic characteristic most affected by the new conditions of arbitrariness is the system's functional differentiation, in the sense of a second-order observation of the system's function.

Any definition of the function of environmental law would have to commence from a basis of contextualisation of the system. Contextualisation does not offer a way out of autopoietic isolation, but an increase of the latter

---

125 Luhmann, 1994c.
126 Luhmann, 2004.
127 Luhmann, 2004:147.
128 'There is no longstanding social consensus about the central question of modern environmentalism – the "correct" human stewardship – relationship to the natural world. Thus, any new relationship has to be created not recognized' (Tarlock, 2004:223).
129 See below, Chapter 4, specifically on the precautionary principle as presumption in the face of uncertainty.
130 Luhmann, 1989b: 63–75; also 1993a.

in the form of inclusion of geographical and historical considerations in the topological self-description of the system. The environmental legal system is required to be spatio-temporally aware through a self-description of adequate environmental sensitivity – the epithet is intended in its *double entendre*. Autopoietically speaking, the environmental legal system is required to include its environment not only as a terrain for cognitive expeditions, an active consideration in the allocation of programme values and a factor of decision making, but also, significantly, as a source of complexity whose varying spatio-temporal parameters will feature decisively in the evolution of the system. In other words, from an autopoietic point of view the environment of the system is the 'object' of the system. Such an inversion is autopoietically feasible only if one does not wish to replace difference with unity. This means that the environment of the system can never become 'one with' the system, but must always retain its incommunicable distance, while, and this is where the difficulty lies, opening up to its potential identity with its environment. For this to happen, however, two concepts that feature prominently in Luhmann will have to be revisited. The first concept is that of society as the ultimate and all-inclusive supra-system (a discussion on which can be found in Chapter 2).

The other concept that potentially constitutes a problem for environmental considerations is *communication*. As said earlier, communication for Luhmann is the 'atom' of any social system.[131] Communication, together with its binary opposite *perception*, is used by Luhmann to distinguish social from conscious systems as their respective units of operation.[132] In describing communication, Luhmann employs the concept of *information* as inherited by general systems theory but significantly altered. Thus, whereas information in general systems theory is what flows freely between system and environment, autopoietic information is constructed internally in the system.[133] Information is one of the selections performed by the system during the operation of communication, the other two being *utterance* and *understanding*.[134] The three selections that constitute communication come about through a reversal of the usual mode of addressing. In autopoietic communication the addressee (significantly: ego) 'controls' the sender (significantly: alter) through anticipation (expectation) of utterance.[135] Ego receives the distinction

---

131 With communication, Luhmann departs from action theory and attempts to readdress the problems that the action-approach has brought to sociology. As Christodoulidis, 1998:76, puts it, 'these are not mere problems that are avoided in the shift from action to communication, but are in fact brought back into sociology as questions that enrich sociological inquiry'.
132 Luhmann, 1985a; 2000a.
133 Luhmann, 1995a; Teubner, 1993.
134 Luhmann, 1989b:143.
135 Luhmann, 1995a:141*ff*.

alter has already achieved between information (the internally produced external) and utterance (the reformulation of information into something linguistically intelligible). If this operation is successful, understanding (the third selection of the unity of communication) will have been achieved. This would mean that ego has drawn the distinction between meaningful and meaningless, at the same time marking the former. The problem with this description of communication is that it presupposes sender, expression, and success in the sense of understanding. All three clash with the afore-mentioned peculiarities of environmental law (although arguably not with the general legal system) and with what has been described so far as the lack of directionality in autopoiesis. While it is true that this directionality is reversed through the use of anticipation (a typical Husserlian internalisa-tion), the fact that communication cannot take place unaddressed because it would lack understanding, means that communication in its tripartite unity is essentially a dynamic form of connection, however internally constructed, which excludes the systemic accommodation of what Smith, in his exhaustive survey of uncertainty calls *meta-ignorance*, namely an awareness of ignor-ance.[136] This is arguably the thorniest point of an autopoietic description of environmental law, and one to which most of this book is dedicated. The issue will be variously put as 'ignorance', 'unutterance', 'porosity', 'memory of forgetting', and of course, as 'paradox', all of which refer to the limits and limitations both of environmental law and of autopoietic theory.

III THE ENVIRONMENT OF ENVIRONMENTAL LAW

As it stands presently, the system introduces the environment in the system through its external reference. Being the unmarked side of the form, external reference operates as a filter for environmental noise, namely the irreducible complexity of the environment. For the benefit of its autopoiesis, the system is *forced* to mark its self-reference,[137] as well as exclude whatever disruptive effects the acknowledgment of mis-correspondence between self- and external reference may have. As a result, the environment of environmental law in its irreducible complexity remains *absent* from the system.

The environment already within the environmental legal system is a sorry affair of domesticated complexity, a functionally indifferent functionalism that cannot properly impregnate the system with the force of exteriority that the environment of environmental law would otherwise. What I am

---

136 Smith, 1989:6.
137 Luhmann, 2004:118, accepts that the system has the ability to change 'leadership', as he writes, and choose external over self-reference, by referring to its programmes rather than its codes. This, however, reinforces the case of *differend* as explained in the following paragraphs.

proposing here is not for environmental law to be torn apart by its environment. Despite its radical exteriority, the environment is not a 'force that
destroys', as Deleuze and Guattari would say.[138] On the contrary, I am
arguing for a resemiologisation of absence in a way that the environment
remains absent and undomesticated by the system as an irreducible block of
ignorance constantly irritating the system's operations from within.

A more abstract formulation of the problem would refer to distinctions
with built-in asymmetry, namely distinctions whose one side dominates the
distinction.[139] This hierarchical technique of drawing distinctions results in
an anticipated prioritisation of the one side, which is the very side from which
the distinction is being drawn. Colonisation of the other side is momentarily
avoided since the impression is one of a form, namely a contingent co-
presence of both sides where no predetermined starting or finishing point
exists. But the distinction is made on the terrain of the one who draws it.
Jean-François Lyotard puts it memorably when he talks about his concept of
the *differend*: 'A case of differend between two parties takes place when the
"regulation" of the conflict is done in the idiom of one of the parties,
while the wrong suffered by the other is not signified in that idiom.'[140] Even
before the conflict, the wrong suffered, or indeed the language, comes the
choice of language, and that, in the case of *differend*, is reserved for the
one whose language is chosen. When earlier I referred to Luhmann's bias
for the system over the environment (arguably an understandable sociological, but not philosophical, bias), I implied precisely this distinction, which
is performed from the structurally predetermined space of the as yet imma-
terialised distinction. An inversion is not always desired, easy or indeed possible. The environment remains irreducible in its complexity, and the act of
distinction will begin from the system – however, such distinction must be
careful to avoid colonisation of its *sine qua non*, and render instead the other
side visible in its luminosity as ignorance.

Whichever way is chosen, the asymmetry between system and environment
must be maintained (indeed, there is no alternative) for cognitive reasons. If
the environment of the system is the domain of cognitive experiences of the
system in accordance with systemic cognitive openness, its complexity, always
greater than that of the system, guarantees systemic evolution. This takes
place, not so much (or not only) on account of the cognitive ability of the
system, nor just because of the aggressive irritation of systemic operations by

---

138 Deleuze and Guattari, 1987:377.
139 Something of which Luhmann, 2002c:39, criticises Husserl when the latter does not
     acknowledge that philosophy should abandon the transcendental quest and look for ways in
     which to reconcile itself with its own contingency. Regardless of whether philosophy actu-
     ally does this or not, it is interesting to observe the distinctions of a philosophically-
     informed sociologist.
140 Lyotard, 1988:9.

the environment; rather, it occurs precisely because of the impossibility of point-for-point correspondence between the two sides of the form. Such asymmetry between system and environment is the necessary condition for cognitive change. This asymmetry is reproduced in the system whenever the latter has to deal with a mismatch between self- and external reference. This mechanism Luhmann calls *re-entry*,[141] in reference to the reproduction of the difference between system and environment within the system itself, which checks itself against its environment, thereby reinforcing its identity. Re-entry is simply another expression of the inevitable asymmetry between system and environment. For this, any illusion of absolute domestication of the environment should be discredited, not only as colonising, but also as self-destructive. This reveals nothing more than the fact that the environment is to remain absent, no matter how eager the cognitive talent of the system may be. The question is, thus, how to conceptualise such an absence.

To that effect, the most pertinent issue is to attempt a definition of the environment of environmental law. If one were to push aside the inherent absurdity of the task – namely the simplification of the irreducible complexity – one would end up with a duplication of the idiosyncratic nature of environmental law. Like several good psychological thrillers, where the detective discovers that he himself has been doing the murders, environmental law lurks on the other side of the distinction as the undomesticated ignorance of the environment. The relation between system and its environment is one of paradoxical continuum/rupture, where the environment finds itself both outside and inside; however, its internal apparition is just a ghost, an absence which ruptures the cognitive continuum. It is the very idiosyncracies of environmental law that inform the other side of its boundaries, a veritable thematisation of its environment in a twist of self-repression. To put it conversely, the object of environmental law is its environment. It is tempting to dwell on the obvious language play, but there are also other ways, possibly less linguistic, to make the point.

One of them is to establish what there is in the environment of the system. Outside their systems and amidst an environment of ecological determination, Luhmann and Lyotard met and talked. Lyotard recounts: 'It was possible for us to form a small common front against the waves of ecologist eloquence. A two-sided front. There is no Nature, no *Umwelt*, external to the system, he explained. And I added: of course, but there remains an *oikos*, the secret sharer [*hôte*] to which each singularity is hostage.'[142] This exchange is the perfect expression of a form ('a two-sided front'), inoperable in its indistinguishability. For Luhmann, there is no transcendental unity, and the system's boundaries are its limits. But for Lyotard there is an absent

---

141 Luhmann, 1993c.
142 Lyotard, 1993:81.

exteriority that reverberates with an ethical call for the system. The exteriority cannot represent unity, but an interruption of the absence of alter, a vociferous re-entry within the self of the difference between self and other as a memento of exclusion. The other side of the system is nothing but the interruption of its unity with the world, which can never be achieved with external reference, re-entry, or even inclusion. The rupture must remain, since there is nothing to connect with the system: there is no Nature, no *Umwelt*. The system floats in an opaque environment. The boundary separating the two is the unresolvable schism of the form – and one needs to start from somewhere, draw a distinction and call it the first distinction. A form is paradox at its most paralysing, so one starts, and starts always with the system – even if after the incision, like Lévinas, one concedes priority to the absolute alterity.[143] The distinction that produces the *differend* is always first, whatever may be attributed to the other side once the other side has been designated as other. Just as a paradox, the differend cannot be solved. It remains a trigger of ignorance that marks the limit of systemic knowledge, but from within: 'not an *Umwelt* at all, but . . . in the core of the apparatus. We have to imagine an apparatus inhabited by a sort of guest, not a ghost, but an ignored guest who produced some trouble, and people look to the outside in order to find out the external cause of the trouble', writes Lyotard.[144] Although invited as a guest, the environment remains absent, ignored, invisible to the system, because the system is at a loss as to what to do with this guest. But this is exactly what is supposed to happen: this description, miraculously, is also the prescription. The environment must remain absent within. But its absence should be sought.

Bruno Latour, arguably the most interesting contemporary ecological political thinker, puts the same schism (and its need for perpetuation) from his distinctly non-philosophical, non-sociological, post-ecological point of view: nature and society 'were constituted for their mutual paralysis, thus clearly cannot be brought together without further ado'.[145] Nature and society are to be kept different, yet one is to internalise the other, not just as a deposit of 'pollution',[146] but as a counterpart that remains alter and leads the internaliser towards the internalised exteriority: 'the concern for the environment begins at the moment when there is no more environment, no zone of reality in which we could casually rid ourselves of the consequences of human, political, individual and economic life.'[147] It is when the environment

143 Lévinas, 1979. One could say that in Lévinas, the priority belongs not to the other, but to the incision as the ultimate manifestation of an absent self facing the other.
144 Lyotard, 1993:100.
145 Latour, 2004a:57.
146 Latour, 2004a:57: 'the prison of the social world would make it possible to subject . . . nature to a permanent threat of pollution by violence.'
147 Latour, 2004a:58.

departs from the scene as the depository of systemic debris and becomes absent that the system turns towards it, its absence, its trace of presence within, and worries. Latour urges us to 'let go of nature'[148] without the fear of losing the exteriority that guarantees the schism, because only then will the system be able to start looking for the environment where it really is to be found: within the system.[149]

So, if nature is no more,[150] if there is no *Umwelt* outside the system, and if exteriority remains absent within, then where is environmental law supposed to look for itself, its 'object', its function, its identity? In a post-ecological time, when humanism cannot be relied upon to bring any change,[151] when ethics have been superseded by scientific discourse,[152] and when even functional differentiation's ability to deal with new risks and ecological crises is being doubted,[153] the system has neither a solid inside nor an alluring outside to look for identity. Within the system's boundaries, the text is infinite but the 'injected' noise, in the sense of environmental meaninglessness, remains substantially filtered by systemic communication.[154] Even in William Rasch's beautiful dual formulation of noise as 'inherently ambiguous, neither desirable, nor undesirable in and of itself',[155] noise remains nested in the environmental complexity and outside the systemic boundaries. Noise could only be 'injected' in the system as order (or indeed orderable noise), but then it would cease to be noise.

The outside, on the other hand, remains a site of reconnoitring, a source of information, the place of potentiality of meaning. But, as Luhmann vertically announces, in no way can the outside be the vanishing point of unity, or the locus of transcendental necessity.[156] Knowledge in autopoiesis can only be circular, self-referential, and, to begin with, asymmetrical. If the environment is always more complex than the system, then the impossibility of point-for-point correspondence between them means that the system is irritated into cognitive evolution precisely because of the uncontainable

---

148 Latour, 2004a: Chapter 1; He is not alone: see, e.g., Eder, 1996; Elliot, 1997; Haraway, 2004.

149 Latour proceeds to a domestication and socialisation of nature which is certainly not what I am suggesting here. Still, his project is one of political ecology, and in view of his suggestions for the collective, socialisation is perhaps understandable.

150 *The Death of Nature* has been famously announced by Carolyn Merchant, 1980, followed by *The End of Nature* by Bill McKibben, 1990. See also Soper, 1995, and, from a phenomenological point of internalisation, Guzzoni, 1996.

151 Blühdorn, 2000.

152 Tarlock, 2004.

153 Luhmann, 2004:156.

154 'texts remain readable to the extent that they remain "infinite" . . . and texts remain infinite to the extent that they remain complex, where complexity registers the "incompleteness" of information' (Rasch, 2000:55).

155 Rasch, 2000:61.

156 Luhmann, 2002c.

complexity outside. Asymmetry reinforces circularity through the generation of cognitive opportunities. Asymmetry guarantees the unity of the system by reinforcing its selectivity. Asymmetry throws into relief the outside as the perpetually incommunicable space of critique. The outside has not been lost, as Adorno and Horkheimer feared.[157] It remains the place of mourning,[158] where ideas about justice, ethics, openness and power circulate freely – except that these are never picked by the system, and when they are, they are 'ordered', phrased in utterances at which the system nods and moves on.

If inside is order and outside is disorder, environmental law, qualitatively more than other legal disciplines, is required to perform a balancing act in order to compensate the influx of environmental uncertainty and the systemic allegiance to legal order. This act can only take place on the boundary between system and environment, on the palpitating line severing self- and external reference.[159] Environmental law is looking right on this line, now here now there, for its identity. The implication for the theory is a departure from the kind of distinctions with built-in asymmetry which give rise to situations of Lyotardian *differend*. The implication for the boundary of the system is that, now more than ever, it is conceptualised as unpredictably fluctuating. The implication for the system is that environmental law buys time, as it were, in order to find ways of accepting environmental absence within, without either exposing itself to the draughts of exteriority, or immuring itself to the complacency of the general legal system. Finally, the implication for the environment is that it remains absent within the system, unfamiliarised and undomesticated, but with a purchase that it would not have, had the distinction been based on the system itself. This is not typical external reference but, to push the terminology further, the external side of external reference. This absent environment cannot be domesticated by any external reference. The system becomes the other of the other, and in that movement, the system allows itself to question itself.

Thus, to return to Lyotard, 'the secret sharer [*hôte*] to which each singularity is hostage' is not Nature, ethics, justice, or indeed the law. The host is to be found on the very boundary between the nothing outside and the nothing inside, on the very ambiguity between self and other, on the fissures of the definition. The nothingness of the environment is placed within the system as

---

157 Adorno and Horkheimer, 1997.
158 Rose, 1996.
159 But also in the relation between *structure* and *autopoiesis*, which very roughly equates to environmental unpredictability versus systemic continuation. Structure is the medium on which autopoiesis takes place, and its main function is to facilitate autopoiesis despite unpredictability. However, even this division, an internal functional division of the system, can be dealt with as self- and external reference, especially considering its provenance, which is the distinction between autopoiesis and medium in biological autopoiesis. See Luhmann, 1995a, Chapter 8.

a hole of incommunicability, a memento of uncertainty and fear, but wrapped in a cloak of familiarity (for otherwise, it could not be in the system) which registers as meta-ignorance: I know that I do not know. Simultaneously present inside and outside, the nothing of ignorance spans the intentionality of system and environment and creates an absence that remains unclassifiable yet always relevant, a nagging presence of irreducible complexity. The system is the environment and the environment is the system, but the two can never meet except in an asymmetrical absence.

IV IN ASYMMETRY

Autopoiesis starts with a Gordian schism, the distinction that marks the form. But before that, there is paradox, constantly re-emerging in the form of reciprocal appellation of the other side. Externally, the system differentiates between itself and its environment by defining itself negatively: I am what I am not. Internally, the values of the binarism are never permanently attributed to anything: what I am not, I can always become. At the same time, the system can never become something different to what it already is: being and becoming are fused into one unity of distinction. The environment of the system, although unobservable, is the arsenal of potentialities of becoming – and being – for the system. This happens because the system remains cognitively open to its environment, but always on the basis of the projection of already existing processes onto themselves. Every novelty has already been there.

But has it? Confronted with ecological problems of unprecedented urgency, expanse and velocity, society is trying to connect its communications with those of environmental changes and prevent the probability of the ultimate (non-)novelty: that of the disappearance of the system. What is slowly happening in environmental law is the re-emergence of the originary paradox, but seen negatively: thus, on either side of the schism there is an absence. Outside, there is no longer nature, ethics, humanism. Inside, there is an external reference to an environment that is not representative of what is happening outside – of all the absence around. But in its absence, the environment is both outside and inside, linked intentionally and at the same kept perennially different.

This arrangement remains asymmetrical. De-paradoxification comes with asymmetries. Asymmetry is the result of selection. Asymmetry is difference. Any introduction of the environment within the system is an affront to asymmetry because it replaces difference with unity: system and environment as one. This, however, only leads to further distinctions because only thus can there be observation. For the whole to exist, it needs to divide itself into the part that observes and the part that is observed.[160] If the system embraces the

---

160 Spencer Brown, 1969.

environment, the resulting holism will only be temporary and of dubious utility. On the other hand, environmental law cannot be without its environment, however constructed. It cannot carry on reverberating on the resonance that ecological issues create within society. As a matter of identity, environmental law is reaching out of society and attempts to accommodate, not resonance, but the environment as such. But this can happen only asymmetrically: the environment is to remain environment, that is, non-selected, and as such will it be seen by the system. Not as part of what the system identifies (with), but as part of the system that remains unknown to the system. Asymmetry within unity; in other words, internal difference.

The paradox can be dealt with only when accepted. Acceptance, however, is self-description. In order to accept the paradox of its included environment, environmental law may need to expose itself to unanticipated probabilities, which will cause it to stretch its cognitive limits and lose some of its protective armour. *En route*, operations and concepts may need to be redefined and the resonance of such redefinitions will be felt predominantly within the environmental legal system. If this takes place and environmental law does not come out altered beyond recognition, it can only mean that the system is equipped to become what it is to become. One can only describe such possibilities with all the risks of wrong predictions and the inconvenience that the impossibility of impositions entails.

Chapter 2

The city
Autopoiesis, society, reality

In the previous chapter, a conceptualisation of the environment within the system has been performed in the form of asymmetry. The systemic environment of the environmental legal system has been found in a simultaneous relation of continuum and rupture with the system, thereby instituting a presence within the system, which, however remains inaccessible to the system itself. Such presence/absence is further examined in this chapter, this time in the context of the city. The city is deemed an appropriate terrain on which to proceed with an elaboration of the above form for various reasons: the institutionalised absence of the natural environment from the city; a replacement of the traditional definition of environment in the context of the city with human-made and exclusionary constructions; and a relation of ambiguous continuum between the city and its resources, ranging from the parasitic to the conceptually akin. The above will be put on an abstract level as the location of the city on the cusp of continuum/rupture with its environment.

The autopoietic context in which such an analysis takes place is one of resistance to the concept of autopoietic society. In its stead, a diffusion is suggested between a reality that refers to everything, and a utopia that cannot describe anything but itself. The city is found to operate on a chunk of 'reality', a space in-between those two extremes, and at the same time, an entrance hall to the autopoietic definition of the urban system, which, unlike the relatively clear-cut social definition of the environmental legal system, necessarily spans the social and the conscious, without however residing on either. Such a 'suspended' definition delimits a space for an environment that remains absent within, through the reiteration of a form that includes simultaneously continuum and rupture with the system. Thus, the urban environment, the environment of the city, and the city in its environment become a terrain of continuous rupture which radically removes any sense of need of identity, except in the sense of operationality – and a disrupted one at that.

I CITY AND SOCIETY

Locating the city requires a spanning of matter and meaning, corporeality and abstraction, tangibility and inaccessibility. These qualities rest side by side, providing for a shifting ground on which distinctions take place. Autopoiesis is not conceptually equipped to locate the city, although it offers adequate mechanisms for such an endeavour, which, when employed, persistently undercut themselves and their designated function. In this chapter, just as in the previous, I will be using autopoietic tools somewhat despite themselves, in order to locate the city in its form. As is obvious, the insistence upon the word *locate* is an indication of the intention to spatialise meaning within the urban context, and approximate a definition of the city as a terrain of fluctuation between the social (namely, social systems) and the conscious (namely, humans). In this process, the environment is once again present in its absence, as the locus of continuum and rupture with the system.

The most prominent difficulty in locating the city is that, while the city is entirely constituted by signs, it continually distances itself from them. Nobody puts this better than Italo Calvino in his description of the city of Tamara:

> The entrance to the city is through streets full of signs coming out of the walls. The eye meets no things, but shapes of things that imply other things; ... If a building has no signpost or figure on its walls, then its shape and position in the city plan are enough to reveal its function; ... Even the merchandise the salesmen spread on their benches have a value, not in themselves but as signs of other things; ... The stare runs through the streets as though they were written pages: the city dictates everything that you are supposed to think, it makes you repeat its own words, and while you think that you are actually visiting Tamara, you do nothing else but register the names with which the city defines itself and all its spaces.[1]

While organising itself in terms of its signs, the city deliberately distances itself from them, enabling a replication of this organisation in the mind of the observer who happens to move in it, while at the same time disabling any enduring coincidence between signified and signifier. The city is located on the boundary between confluence and conflict, projecting itself on this very pulling and pushing with its signifiers. There is an urban revolt against the city's own textual organisation, thus re-instituting the rupture between its description and its self-description – in other words, its signifiers and signifieds.[2]

The city is always a spanning of a fissure, and in that sense, a precarious projection that tries to accommodate various opposing potentialities. The

---

1 Calvino, 1993:13–14. The translations from the Italian of this and the other two stories that follow in this chapter are mine, so any errors are my responsibility.
2 'If there is a textual system, a theme does not exist. Or if it *does* exist, it will always have been unreadable ... Meaning is nonpresent, or nonidentical, with the text' (Derrida, 1981:250). Indeed, the urban process remains unreadable, but this unreadability in the form of simultaneous identification and distanciation is here operationalised as an urban paradox.

city's actuality is but the fusion of such potentialities in a contingent but probable form, while retaining on the other side of such actuality the distance from its own fusion of potentialities. Thus, the city is projected both by itself and by each observer: 'cities, unlike villages and small towns, are plastic by nature. We mould them in our images: they, in their turn, shape us by the resistance they offer when we try to impose our own personal form on them . . . The city as we might imagine it, the soft city of illusion, aspiration, nightmare, is as real, maybe more real, than the hard city one can locate in maps and statistics, in monographs on urban sociology and demography and architecture.'[3] Soft or hard, the city is a projection of the contingent fusion between a social and a sensory adventure. In these fusions there is no 'thing': 'we classify an environment as a city, and then "reify" that city as a "thing". The notion of "the city", *the city itself*, is a representation.'[4] The city is re-codified every time a visitor enters it, an observer climbs up the tallest turret and stares at it, a planner dreams of redesigning it, an inhabitant gets lost in its streets. Maps do not locate the city. They do not even locate their reader in the city. They can only allow for a precarious location (of the city, of the reader) on the fusion between the social and the conscious (by marking an ethnic, political, religious or other social necessity to project the representation of a specific perception),[5] but without the possibility of fractally changing and crossing from one to the other in order to follow the location of the city.

Thus, habitual metaphors that emphasise constant flow, anarchy, polysemy and mobility in the city, fail to grasp the urban *form*, marking in that way only one of its sides. Likewise, descriptions of the city only from the citizen's point of view, or only from a social institution's point of view, also fail to deliver the subtlety of the form. An adequate description of the city must capture the spanning between flow and order, conscious and social, and dwell on neither but always on the act of *crossing*. It is on this performance of the city, its distanciation from the signifiers while encouraging their crystallisation, that the city can be located. This complex form of being/becoming is in need of epistemological crutches in order to be described, for otherwise the object of description becomes diluted in the very act. This is the reason for which urban analyses focus on 'representations' of cities rather than the city itself.[6] Even when they deal with specific cities, or pragmatic considerations of the urban environment, urban analyses are but epistemological fractions of the urban complexity. Even the choice is uncontainable. A perfunctory list would include mechanical, cinematographic, organic, artistic, literary,

---

3 Raban, 1974:10.
4 Shields, 1996:227, original emphasis.
5 Sousa Santos, 1995.
6 See for example Bell and Haddour, 2000; Westwood and Williams, 1997; Watson and Gibson, 1995; and McAuslan, 1985 and particularly Chapter 9 on the containment of contradiction between legal and illegal city.

and statistic formulations, as seen from social, historical, geographical, human geographical, economic, sexual, political, ecological, sexual, linguistic, demographic, utopian, racial, gender, psychological or architectural points of view: all ways of fragmenting and approaching the urban complexity, for 'precisely because of this complexity, the city can never be wholly fathomed'.[7] Unfathomability is an indication both of the absence of any *über*observer, and of the fact that the city is what there is – and this is said devoid of any vein of urban imperialism. In wishing to observe the city, the observer reinstates the revered role that the urban has acquired in the history of human civilisation,[8] and does little more than express a desire to study the human organisation, without further specifying which aspect of it. In its abundance of symbols, the city, not unlike a map, has itself been converted into a symbol of its featured symbols: '[t]he city has come to be a symbol – maybe even a symptom – of almost every social and cultural process. Cities are certainly concentrations of these processes: the city is often read as the medium through which modernity (and then postmodernity) gets expressed, worked through, concretized.'[9] From Benjamin's *Arcade Project* to Baudrillard's *America* and Davies's Los Angeles, the city has been designated as the decontextualised locus of a historicity and spatiality that typically transcend the pragmatic and operate on the level of 'treacherous metaphors'[10] as representations of society.

This representation of an all-encompassing, all-thematising but elusive signified, positions the city conceptually close to Luhmann's definition of *society*. According to Luhmann, 'society is the all-encompassing social system that includes everything that is social and therefore does not admit a social environment'.[11] For Luhmann, there is a specific formation, which, despite appearances, resists hierarchy:[12] there are social systems, such as law, politics, and so on, functionally differentiated from one another. All social systems are sub-systems of society – the supra-system. As said in the first chapter, society does not include humans or nature, only social systems.[13]

---

7 Amin and Thrift, 2002:92.
8 Mumford, 1961.
9 Bell & Haddour, 2000:1, emphasis omitted.
10 Shields, 1996:229.
11 Luhmann, 1995a:408.
12 Hierarchy is avoided because of the autopoietic turn to a qualifiedly solipsistic system. See Schütz, 1994, and Murphy, 2001, who disputes hierarchy by putting messages from leadership on an equal footing with other messages, although pointing out what he calls 'crypto-normativities'. Also, Murphy, 1997, suggests thinking hierarchy in terms of 'horizontalization', which generally moves along the lines of differentiation.
13 'Nature' is a very inclusive concept indeed for Luhmann: 'Nature extends from biological phenomena, such as events in the human body or in the natural world of plants and animals, to starts, planets, atoms and nuclei, to natural catastrophes such as earthquakes or global warming' (King and Thornhill, 2003:8).

Humans, even in their conscious rather than corporeal manifestations, lie outside society, because, although they operate with meaning, the meaning produced by conscious systems (what humans are in autopoiesis) is different to the social systems' one. Conscious systems operate with perceptions and interactions, whereas social systems operate with communications.[14] The operations of conscious systems are not understood by a social system because they cannot be dealt with on the basis of the systemic code – for example, a perception remains a perception and can never qualify as lawful or unlawful.

The problem is not that perceptions are not included in society.[15] The problem is the need for a concept that would aggregate systems in anything 'more' than their already expansive boundaries. For Luhmann, society is 'the social system whose structure regulates the ultimate and basic reductions to which other social systems can be attached. [It] guarantees for the remaining systems an almost domesticated environment of reduced complexity.'[16] In other words, society is the ultimate environment for social systems, outside which there can be nothing else: 'For [society] there are no environmental contacts on the level of its own functioning . . . It is completely and without exception a closed system.'[17] As a result, society embodies the concept of self-referential closure at its absolute, and can never communicate with its environment but only *about* the environment by way of internally produced communications. Anything meaningful is of society; anything outside society is meaningless, hence incommunicable to society. Society operates with communication in a manner akin to monopoly: 'society carries on communication, and whatever carries on communication is society.'[18] Thus, society is the only source of meaningful communication and its only recipient; anything outside society simply cannot exist for society.[19]

Typically for a self-referential theory, autopoietic society needs to be defined in this way because only thus can systems carry on producing communications and therefore their autopoiesis. In that sense, society is an epistemological necessity of sociology (and this society, an epistemological necessity of autopoiesis), since sociology needs a concept to express 'the unity of the totality . . . of social relations, processes, actions, or communications'.[20]

14 Luhmann, 1995a; 2000a.
15 See below for the brunt of antihumanism.
16 Luhmann, 1985a:104.
17 Luhmann, 1995a:409.
18 Luhmann, 1995a:408.
19 However, two things compromise this position: first, the concept of world society, which is 'the innumerable multiplicity of simultaneously ongoing communications' (Schütz, 1996:283), and second, the interaction systems which run parallel to society and consist of perceptions.
20 Luhmann, 1995a:408.

Luhmann himself acknowledges that his chosen definition of society has been guided by his attempt to avoid the practical difficulties with which social constructivism is associated.[21] But, what is the price for such an epistemological concession? Strictly within the ambits of autopoietic theory (which means without any intention of expressing doubts with regard to any other form of society except for the autopoietic one, namely the totality of communications), autopoietic society risks being a partial description of the way society operates, because of the impression of efficacy of domestication for the various social systems, who in fact are constantly more and more exposed to (internally constructed) gaps of communication. This is not simply the kind of negation already included in society in the form of failed communication,[22] or the necessary immunity with which social systems must armour themselves in relation to problems outside society (that is, conflicts between conscious systems in relation to social systems).[23] The gap of communication which society cannot include within its autopoietic definition goes before and beyond negation, remains unconnected with anything meaningful or meaningless, and resists familiarisation through history, reference or re-entry. Autopoietic society does not accommodate its *aporias*, its foundational, unutterable paradoxes,[24] because the latter put into question society's very boundaries.

To put it more specifically, society is a mode of resistance against self-questioning. Luhmannian society is the womb of selections, in which systems and their environment cohabit. The evolution of systems within society is exposed to the theoretically infinite horizon of probabilities their environment offers – hence, their cognitive openness. An autopoietic system selects its potentialities from the horizon of probabilities, and its evolution is blind; first, because its environment is invisible to the system; and second, because everything is decided internally, not on the basis of some perceived 'need' of the system, but on a contingent basis: '[a]nything is contingent that is neither necessary nor impossible'.[25] Contingency is perhaps the cornerstone of the processes by which autopoietic operations occur, precisely because it guarantees (by its very incapacity to guarantee anything) the unpredictability of the outcome. Contingency conceptually represents the way the horizon of probabilities makes itself 'visible' to the system. Whatever attempts to contain contingency creates a fallacy of a containable horizon, which, nevertheless, appears to the system as the only horizon available. Every concept, however comprehensive, performs a distinction between itself and its environment. If

---

21 See Luhmann, 1985a; 1988a.
22 Luhmann, 1995a:445.
23 Luhmann, 1995a:423.
24 On the concept of unutterable paradox, see below, Chapter 3, and Philippopoulos-Mihalopoulos, 2005b.
25 Luhmann, 1998a:45.

society can have 'no environmental contacts on the level of its own function-ing',[26] it is because the pre-selected availability of probabilities as represented by society, curtail contingency to the point of a partial horizon which can only offer probabilities that can be communicated. Anything else is excluded from society: it becomes excommunicated.

An argument that unsettles the limits of Luhmannian society can be sus-tained even in the face of some of society's mitigating circumstances: it is indeed the case that, according to Luhmann, society is a happenstance, an improbability which happened to happen and continues to happen for as long as it manages to.[27] Society is an aggregation of its systems, and receives its limits from the boundaries of its systems. This means that society exists as nothing more than an absence of environment, a hole in the environmental chaos, a nebulous formation which sustains itself as a bunch of communica-tions huddled together. While this does address the issue of direction (namely, no longer an imposition from society to the system, but the other way round), it reinforces the issue of communications as the sole distinguishing oper-ational unity that characterises observable systems. Even if it is the system that defines society, society lends to the system its epithet, and baptises it with a communicational talent that operates protectively against the disruption of such epithets. At the same time, even if society is simply a representation of the total selection of its sub-systems (in other words, it is not society that delimits the horizon, but the systems themselves), the systemic limitation is reinforced autopoietically, in the sense of superimposition of the *totality* onto totality, thus totalising the cognitive experience within a set of (illusionary) secure exclusion. This means that, although total, society is not a form but a selection. Even as an ondoyant cluster of communications, society remains a limiting concept that operates maternally to her systems. Society converts herself into the garden of toys with which systems can play. The wall around the garden, however, is too high even for the mother herself, and the family ends up isolated in its self-delimited fallacy.

But is this fallacy really self-imposed? Autopoiesis rejects the hierarchical dialectic of whole-parts, and places in its stead the monolectic binarism between system and environment, where one is folded into the other as poten-tiality. Every system is autopoietically perceived as a replication of the greater scheme of potentialities, a fractal shoot palpitating to the rhythms of its environment. Consequently, social systems cannot be responsible for the selections of the totality. It is the totality that selects for itself and from itself, but its selections affect the further selections of the systems it purport-edly includes: the maternal becomes paternalistic in a characteristic fear of losing authority. Society's authority is a very loose authority indeed, a

26 Luhmann, 1995a:409.
27 Luhmann, 1997a.

pseudo-mastery that extends within (in the replication of totality via communication) and shies away from the outside, from a powerful environment that collapses under the weight of its own gravitas. In its crippled, indeed vaporous omnipotence, society rests on its boundaries that include and are determined by what the systems exclude, while never including the absence of exclusion – thus remaining a selection and never re-turning to being a form. And, while there ought to be no hierarchy anywhere to be observed, the act of observation imposes a pseudo-hierarchy between the observable society and the unobservable environment, compromising the subtlety with which autopoiesis should circumvent considerations of hierarchy, with the imposition of an internal hierarchy whose feudalism seems conceptually asphyxiating.

An epiphenomenon of society's epistemological construction is the risk of an epistemological dilution of autopoiesis.[28] The system's autopoiesis is 'purposeless',[29] for according to Maturana and Varela, purposes are not features of the system but belong to the domain of the observer: 'they reflect our considering the . . . system in some encompassing context.'[30] Is a social system placed within society because it has to preserve its societal function? If so, then teleonomy returns forcefully but inappositely. Is society perceived as the supra-system in order for the complexity of the encompassed systems to be reduced? If so, then it belongs to the level of second-order observation and not to the system. The bifurcation between system and observer is one of the most capricious paradoxes of the theory. The introduction of second-order observer is Luhmann's necessary step away from a biological autopoiesis, which took place on the basis of a unidirectional, constructivist observation of the autopoietic cell (whose opening sentence was '[a]nything said is said by an observer'),[31] to a systems theory-informed social autopoiesis, whose second-order observation enables and observes an ambidirectional operation of observation to take place between various systems. While the edifice is justified in view of the sociological distance from the object of study and the demolition of the subject/object binarism, it debilitates the paradigm social autopoiesis specifically attempts to introduce: namely, the shift from external factors to internal operations.

Following the same line of argument, one can criticise the concept of functional differentiation.[32] The opacity of functional differentiation is only visible to a second-order observer and even then not always convincingly. Each system deals with a social communication differently, but this difference

---

28 Or a problem of definition: 'how a subsystem of the social system can itself be autopoietic within the autopoiesis of the entire social system?' (Jacobson, 1989:1650).
29 Maturana and Varela, 1972:85.
30 Maturana and Varela, 1972:85.
31 Maturana and Varela, 1972:9.
32 See Habermas's critique, albeit not for the same reasons, 1996.

is invisible to the system *per se*.[33] Society is a construct of second-order observation, which, however (or precisely because of that), permeates the actual construction and affects first-order observation between systems.[34] The dilemma is formidable. It is difficult to imagine a society without society, or a theory without second-order observation. But is one expected to remain faithful to the theory or to reality? And in the event of a conflict between the two, must one abandon the theory as empirically unfounded?

What appears to be a dilemma is simply another form, one side of which implies the other. As Merleau-Ponty puts it, 'solitude and communication cannot be the two horns of a dilemma, but two "moments" of one phenomenon'.[35] The solitude of the description from within the system (its self-description) cannot operate without the communicative operation of observation of the system by another system (description) in a circuit of second-order observation.[36] Luhmann's pronouncements that the theory talks on behalf of the system are conditioned by the prominence of the second-order observer. It is, however, problematic when theory, via second-order observation, posits concepts that, while seemingly reflecting empirical reality, in fact operate as a hindrance to the theory's connection with it. Society as the limits of observation,[37] with its dominating distinction

33 Another way of looking at functional differentiation could be as facets of the multiplicity of the self (Teubner, 1989), but this again would demand second-order observation (see also Baxter, 1998:2015), and would not succeed in leaving out functional differentiation. The latter could be understood as functional *plurality*, thus avoiding the comparative element of the former. From the second-order point of view, differentiation still exists, only that there can be no credibly communicable guarantee as to the limits of each differentiated system except in the observational 'reality' – see following section.

34 The same can be put from the point of difference between observations and constitutive operations, where the former refer to the ability of a system to observe other systems and itself, while the latter are the autopoietic operations the system performs when constituting its identity, when being and becoming. Communications are about the observations on system and environment, since the only avenue for the system to perceive itself and its environment is through observation. Communications and observations are both systemic operations 'linked internally in a mutually enabling way' (Christodoulidis, 1998:81). However, one thing is the description of an observer of a system, and another the autopoietic operations of a system, which are internally produced, consumed and observed, which do not obey to any teleonomy, and which are not aware of their environment *per se*. The two are not only different in their difference, but also incommensurate. The fact that at some point they seem to merge into undifferentiated identity is only impressionistic: it is the results of observation that may be converted into systemic operations (a manifestation of the cognitive openness of the system) and not the operation of observation itself. See also Teubner, 1993, Chapter 2, and Luhmann, 1998a:18.

35 Merleau-Ponty, 1962:359.

36 Luhmann, 2004.

37 Only social systems are capable of being observed; Luhmann, 1993d:773 puts it wonderfully dryly when he excludes human beings from observation because 'there are too many of them'.

by and for the system (for society is a system after all), its inability to allow systems to expose themselves to extrasocietal ignorance, its own impotence to risk its boundaries in any substantive way: all this poses an important problem of miscorrespondence and unhelpful theoretical reduction between a society that is awakening to its ignorance and attempts to find ways of accommodating it, and an autopoietic society that ignores its ignorance.

A modified understanding of autopoietic organisation is possible, one that returns to the system and its internal operations, while momentarily minimising the luxurious security of the second-order level. The impossibility of focusing on the level of operation is salvaged by two contrasting concepts: its very impossibility, and intentionality. The focus on operation is always compromised by its impossibility – hence its relevance. As a succesful failure, the focus on operation escapes its possibility too, precisely because of the ever-present intentional connection between observed and observer. Thus, its impossibility (as escaped via intentionality) is what makes it a plausible avenue here: through this minimal focus (one could say, lack of focus), absence can be included in the system and remain absence. Avoidance of second-order observation (however illusionary) amounts to lack of attribution, namely connecting strands between what and how. This is wished for here. Unencumbered by the protective atmosphere of a supra-system, and in a self-referential whirlwind that remains incommunicable to any other system, the system returns as the only terrain on which the absence is to be studied: an autopoiesis that demands viewing the phenomenon of society, in its altogether surprising and incidental apparition, from the boundary of a system, indeed from a point of internalisation not only of systemic processes and operations, but also of its aporias: from the very distance between observer and observed. In an autopoiesis that does not employ societal boundaries in order to carry on with its communications, communications are disrupted – as indeed they are – by an environment whose incommunicability guarantees its absence and, at the same time, prohibits its normative colonisation on behalf of the system. In other words, lifting the shield of autopoietic society from autopoiesis exposes the system to the draughts of an unmasticated environment, full of potential communications, but also actual perceptions, fantasies, imperceptions, impossibilities, and other destabilising factors that demand a systemic reference different from the self-referential unity of self- and external reference. Since the gap between the two references grows exponentially, the system readjusts its self-description to unfiltered uncertainties, producing expectations that reveal its systemic limits and limitations to itself. Adequately exposed theoretical premises are necessary for the description of an inadequately compartmentalised environmental uncertainty.

The removal from the theory of the explanatory tool of society does not side with humanist criticisms on the autopoietic exclusion of human beings from society.[38] As already said in the first chapter, this is one of the main areas of misunderstanding when it comes to autopoiesis, in that its non-humanist focus is mistaken for a technocratic, non-humane focus.[39] The accusation is both misplaced and incomprehensible. Except for the fact that Luhmann has often explained that locating humans outside society is a way of recognising relative freedom from specific social positionings,[40] it is remarkable that Luhmann is accused of not espousing something already rather heavily discredited. Humanism is another grand narrative, complex in its ambition and problematic in its scope, which has led to justifications of concepts and acts that could not have been accepted as part of humane (but all too human) behaviour.[41] Thus, the famous 'exclusion' is neither here nor there, and its precise nature has nothing to do with *anti*humanism, but with a subtler relation between communication, on the one hand, and systems of *interaction*, on the other, based on human perceptions. The mechanisms will be explained below, but for the time being it is necessary to note that there is no clear-cut boundary between communication and perception, at least the way criticisms seem to understand it.[42] Thus, even without the societal boundary, systems continue with their communications, human beings continue with their perceptions, and the already existing mechanisms of cross-fertilisation still hold. The reason for this is that, with the removal of society's boundaries, the question has moved from the difference between society and its environment, to the boundary between the system and its observer. It is on this very boundary that the present discussion projects its epistemological object. With this, autopoiesis returns to being the art of distinction, and not necessarily of definition. And while distinction may lead to definition, the impassionate bleeding of the distinction discourages any stability that may lead to conceptual stagnation of the theory, and takes into consideration an environmental complexity whose limits are well advanced into the trenches of the system.

In turning to the city, I take my cue from Tim Murphy who suggests that 'instead of regarding "society" as some encompassing system, in relation to which these others must be termed sub-systems, and merely reversing the old hierarchical scheme by switching into the driving seat the object rather than the subject, the ruled rather than the ruler, we need a cooler, more banal

---

38 See above, Chapter 1.
39 E.g., Bankowski, 1996, and to a lesser extent, Wolfe, 1992; see Luhmann's retort in 1995a:210–13.
40 Luhmann, 1995d.
41 A new humanism is being suggested, which destabilises the blindfolded trust in human nature of the original strands of humanism, thereby allowing for new entries. See especially the feminist lliterature of Haraway, 2004; Braidotti, 2006; from a different point of view Douzinas and Gearey, 2005.
42 See Luhmann, 1995a, Chapter 10, as well as 1994c.

vision . . .'.[43] Although any city contains several rather uncool corners, it is certainly more banal than the unity of totality of communications, which society is. I turn to the city, not so much as an epistemological supplant of society, but mainly as an indication of an absent society, both inside and outside the city. To put it differently, this is not the place to construct a sociological theory, which as such would probably need a demarcating concept of society. My attempt at conceptualising an autopoietic understanding of the urban environment does not even necessitate a concept of *koinonía politiké* in the Aristotelian sense of political and moral autarchy. Instead, the present focus is what has been described at the beginning of this section as the most prominent difficulty in locating the city: the distance between the 'thing' and its representations, which can also be put schematically as the difference between self-description and descriptions, or system and observer. Hence, the location of the city is to be found in neither of these (since the 'thing' does not exist, and its representations are all-encompassing) but on the very distance between these two things. The city is a fractal reiteration of the boundary between perception and communication, self-description and descriptions, sensory apprehension and representation, corporeality and spatial co-existence, monadology and nomadology, system and its observer: in short, a form that relentlessly repeats itself without ever losing sight of its original paradox. Through a catenation of autopoietic events, the city continues its mutual undercutting with its environment while reserving a place for undomesticated ignorance. The tool for such continuation is necessarily one that can fluctuate along the distance between thing and representation, without pinning down the city in either of those exclusively, nor, however, relinquishing an observational desire towards the seemingly impossible. This tool I call 'reality' as a continuation of the paradox between an all-knowing reality and an all-thematising utopia.

II REALITY, 'REALITY' AND UTOPIA

A certain 'reality' of a city is recreated by every observer with every description. Apart from being numerous, urban representations are also inaccessible to one another because of the systemic boundaries which operate as conceptual barriers between the various representations. Indeed, there can be nothing collective about a description – whether its provenance is social or conscious. To quote Calvino again, 'whichever country my words evoke, you will see it from an observatory identical to the one you are in now'.[44] However faithfully Marco Polo describes his discovered city to the Khan, and however

---

43 Murphy, 1997:174, with different results to the ones here.
44 Calvino, 1993:25.

attentively the latter listens to the words of Marco Polo, there will always be two cities: the traveller's and the listener's. Every description of a city is utopian in its attempt to communicate and construct a commonality of descriptions. Or, to reverse the definition, utopia is the operation of linking an urban representation with other urban representations.[45]

The above can be better explained if one considers that utopia, while firmly rooted in the subjective description of an observer, attempts to provide for an ideal that purportedly applies to everyone. Utopia is typically a 'one-man dream',[46] a personal, egocentric voyage of description of the ideal urban society in the folds of one's self-indulgence. The voyage to utopia is a soliloquising delirium on the surface of which the utopist is mirrored. Utopias are like dreams: 'we can dream whatever our imagination leads us to, but even the most unexpected dream is an omen that hides a desire, or its opposite, a fear.'[47] Utopias criticise reality through the means of an imaginary elsewhere, which is perceived, at least by the utopist,[48] to be better than the actual 'here': 'utopia denotes an elaborate vision of the "good life" in a perfect society which is viewed as an integrated totality.'[49] Indeed, the utopian perfect order, however solipsistic and free from anything haphazard, accidental or ambivalent, is to be appreciated by the entire society.[50] Thus, even if utopia originally were to be a 'one-man dream', its purported addressee is the whole world. Utopia describes itself as a society ideal for everyone. The latter is arguably what lends utopias their utopianism: despite the fact that utopia remains of a specific system, it embodies an unrealisable attempt to converge the systemically isolated – or else, the descriptions of the observers in an all-informing *über*observation of the highest second-order.

Accordingly, utopia is not the ideal society. Utopia is the representation of the ideal of each observer. But precisely because the societal ideal includes and presupposes a space for alterity, utopia is also a consciously failed attempt to bring these ideals together. Utopia's inherent failure folds within utopia itself in the shape of fissures, of spaces where impossibility is recognised; and precisely because of this recognition, utopia can carry on 'existing'. These fissures are reflected etymologically: the word 'utopia' means the non-place (*ou-topos*), the place that does not exist. To discover utopia would be to discover that utopia exists somewhere, which fatuously leads to the proposition that utopia cannot exist as an existing place: it ceases being utopia. To not discover utopia, on the other hand, to resign from the quest of the personal ideal, is equally fatuously echoing the 'death of

45 Further see Philippopoulos-Mihalopoulos, 2005a.
46 Manuel and Manuel, 1979.
47 Calvino, 1993:44.
48 And its editor, since, according to the Manuels, 1966:4, every utopia is at least a *folie à deux*.
49 Goodwin and Taylor, 1982:16, my emphasis.
50 Bauman, 1998, 2003b, 2005a.

Utopias',[51] whereby the negation embodied in utopia is hurled into oblivion, and the voyagers are advised not to travel. It seems that the final negation of utopia would be its existence – or, indeed, reality.

The connection between utopia and the city is still relevant. Utopia is the form a city would have taken, had communication between systems (with a continuing respect for boundaries) been feasible.[52] Utopia remains an elusive concept, and, in this respect, not different from reality. Utopia and reality represent the two extremes of the urban spectrum. While both are equally inaccessible, they are also diametrically different. *Whereas reality should be understood as the collective articulation of the absence of descriptions, utopia is the individual (namely, of each system) projection of a collective articulation of a commonality of descriptions.* The latter means that, not only does the utopist project her own ideal city, but she also presumes (in a state of naïve isolation) that this city is ideal for everyone, precisely because of the presumed commonality of descriptions. The two opposing sides (reality and utopia) are in fact complementary in their attempt to describe the world: utopia starts from the systemic, individual level of description, while reality comes from the vantage point of no-description, the theological point of totality, or indeed the absolute point of second-order observation.[53] The form utopia/reality I call 'reality' and I further operationalise it by marking it, not on the side of reality (second-order theoretical observations), but on the side of utopia (observation from within, without the distance of second-order). Thus, 'reality' is how the form utopia/reality operates when marked on the side of utopia (but with reality always implied and vice versa), and can be defined as the pragmatically operable state of systemic co-existence as perceived by each system that partakes of that co-existence. 'Reality' in this sense expresses the evolutionary *surprise* of communication in view of the incommunicability of the systems – a concept that comes close to the fortuitous improbability of the emergence of Luhmannian society, but without either the limiting effect of division between communication and perception ('reality' can be observed by any observer, be this social or conscious), or the lack of awareness of (in)communicability. 'Reality', in other words, is utopia which, precisely because of its undercutting with reality, has internalised its negation, its fissural topology, and continues in a state of astonishment at how, regardless (or on account) of all the fissures and incommunicability, 'reality' is still the one thing capable of being communicated.

'Reality' is also reminiscent of Maturana and Varela's description of the

---

51 Manuel and Manuel, 1979; Lowenthal, 1987.
52 Even structural coupling, below Chapter 3, cannot enable transplant of forms.
53 A second-order observation which would not generate a blind spot. The result of such an observation would be totally incommunicable precisely because of the transcendence of itself.

relation between construction and reality. In biological autopoiesis, when an observer observes, he establishes axioms: '[f]or the observer an entity is an entity when he can describe it.'[54] Although Maturana and Varela endow the observer with such a divine power, they do not define the canvas from which the observer selects: is it just hidden reality or the abyss of non-existence? 'The phenomenon of knowing cannot be taken as though there were "facts" or objects out there that we grasp and store in our head. The experience of anything out there is validated by the human structure, which makes possible "the thing" that arises in the description.'[55] With this lateral reference to Husserl's intentionality, biological autopoiesis confirms that the conditions of objective knowledge reside in the knowing subject.[56] The emphasis is therefore shifted from reality to the production of 'reality', namely the Husserlian *noesis*: the act of perception of the intentional object (the *noema*). Intentionality brings forth 'reality', disregarding the need to determine whether there exists a reality or a utopia from which the observer is supposed to select (it suffices if the object is constructed intentionally). Both 'reality' and intentionality are operations of the observer, tools and constructions that link the extremes of *über*description (second-order reality) and utopia (internalised belief in communicability). Thus, the observer does not create the world but *their* world, which is the world at large 'brought forth' by the observer.[57]

Luhmann does not employ the above terms. In his comments on Husserlian phenomenology, Luhmann equates noesis and noema with self- and external reference respectively.[58] In that sense, the constitution of the world is a result of the permanent oscillation between the two kinds of reference, just as phenomenological description is the description of the correlation between noesis and noema. Accordingly, intentionality is 'the positing of difference, the drawing of a distinction'.[59] While typical, the equation of intentionality with distinction is, as it were, one side, and for me, the unmarked side of the form. Intentionality's hermeneutical value is precisely its location before (or beyond) distinction, since it *connects by internalising* inside and outside (and this even despite the later Husserl). By understanding it as distinction, intentionality enters the discourse of identity/difference and entails marking.[60]

54 Maturana and Varela, 1972:8.
55 Maturana and Varela, 1992:26. Biological autopoiesis offers no answer as to whether reality exists objectively but is perceived subjectively or merely subjectively existing, although the fact that on the same page as the above extract the dizzyingly circular picture by M.C. Escher of two hands drawing each other features prominently, is indicative.
56 Husserl, 1970a.
57 Maturana and Varela, 1992:27; see also Luhmann, 2004.
58 Luhmann, 2002c.
59 Luhmann, 2002c:45.
60 And marking there is, since, according to Luhmann, all reference is self-reference – see above, Chapter 1.

Allowing it to be conceptualised as the preposition 'of' that connects consciousness and object, intentionality retains the paradox of a form and lends itself as the tool through which to apprehend the world. In that sense, 'reality' is pure intentionality between an observer and his world, bathed in the astonishment that the two are, counter-intuitively, connected despite their severance.

To this, Luhmann arrives from the opposite side: rather than say that reality (in the sense of 'reality', i.e., the intentional connection between observer and object) is present in astonishment, he considers it an illusion, which, however needs to be saved. In what seems like an act of tolerance, he conceptualises it as a necessary solution to the unavoidable inconsistency problems, an 'insight' that compares and links difference,[61] and allows one to move from one construction to another especially in the context of a 'polycentric, polycontextual system' like society.[62] Reality emerges when the observer is faced with inconsistencies, paradoxes and operational contradictions. The observer imposes her 'insight' on the brouhaha of difference and makes sense of it. This 'insight', however, remains unobservable, a tool that makes things appear right without itself appearing anywhere. Indeed, elsewhere Luhmann has defined reality as 'what one does not perceive when one perceives it'.[63] This is Luhmann's way of emphasising reality's relation to the paradox. Indeed, reality is 'the correlate of the paradox of the unity of self-reference and hetero-reference ... And, like a paradox, reality requires "unfolding". It is only an aid for reaching one construction from another.'[64] This boundary-like description of reality accords with the present analysis of 'reality', except that the process is inverted: while Luhmann starts with difference and accepts the necessary imposition of reality, I start with the observation of possibility of communicable co-existence ('reality') as marked by a utopian internalisation of absence. 'Reality', in other words, begins from the paradoxical internalisation/thematisation of a utopian possibility of the world, which, on account of being utopian, excludes itself from ever being materialised. To put it differently, its 'unfolding' is a negative one, with its marking resting uneasily on the environment rather than the system. While it is important to consider 'reality' a tool of 'connection', it is also important to allow its utopian marking to carry it towards an internalisation of the absence of topos, of the knowledge that, even though communication is only an illusion constructed internally by the system, it is still happening and can carry on regardless of its aporias. In that sense, 'reality' is Luhmann's reality but comfortable with the paradox, without the desire of unfolding it.

61 As Luhmann, 1995c, clarified in a discussion with Katherine Hayles.
62 Luhmann, 2002c:52.
63 Luhmann, 1990b:76.
64 Luhmann, 2002c:52.

Arguably, the greatest paradox of 'reality' is the maintenance of the form utopia/reality, despite itself being marked by utopia. This is because utopia is absence, the negation of marking, the side of the form never to be observed, and as such it can never be marked without relinquishing its marked topology to the other side, namely reality. Utopia is a nagging absence of ideality, a non-topos never to be reached, always to be disarmed within its form. Even when marked, utopia is nothing more than like the dot of a beacon on a radar screen. For, as Lewis Mumford puts it, utopias are indeed 'the only possible beacon upon the uncharted seas of the distant future'.[65] Even so: nobody would want to fall on the beacon, although everyone should see it.

The term finds its prime application in the city. Urban 'reality' is what an observer sees when looking at a city as an articulation of systemic co-existence, which prima facie would not have been operable, but somehow is. This paradox translates into astonishment (hence its comparability with Luhmannian society), but is not being unfolded: 'reality' remains intentional, imposes no limitation on the kinds of operations used in its construction, and incorporates the paradox of utopia and reality in its unmarkable, *différant* form. Urban 'reality' expresses the constant astonishment at the uncompromised revelation of the paradox, which is attributed to the autopoietic isolation of a system from its environment, the lack of an over-arching system which would create a domesticated environment of reduced complexity, and the fact that cities are actually perceived by observers and that there is a commonality of descriptions. Urban 'reality' is the embodiment of a utopia that retains the paradox between itself and reality. So, whether society exists or not is not relevant. The 'real' presence of society constitutes the fulcrum between the symmetry of the extremes of reality and utopia, while optimistically marked on the side of utopia. This is perhaps the crucial difference between Luhmann's tools and 'reality': 'reality' is asymmetrical and itself terminally undermining of its asymmetry. It is marked on the side of utopia, precisely because it has matured to the possibility of communicable co-existence and has learned to accept the surprise as a paradox, but its marking is, as Rilke has put it, 'never Nowhere without the No'.[66] This presence on the basis of absence is what helps 'reality' evade the paradox of the absolute symmetry between the extremes – something which the Luhmannian society with its overarching, all-inclusive quality fails to achieve – while, at the same time, pulling towards a haunting absence of potentiality.

Urban 'reality' is a perilous oscillation between the absence of topos and the totality of a theological unity. Located between the two positions, urban 'reality' is a suitable tool with which to describe the city, in view of the latter's

---

65 Mumford, 1923:xii.
66 'Immer ist es Welt und niemals Nirgends ohne Nicht' (Rilke, *The Eight Duino Elegy*, 1995:377).

formation as the difference between the (non-existent) 'thing' and its (all-encompassing) representations. 'Reality's' nuanced marking on the side of utopian descriptions is not a gesture of transcendence, but one of constant self-annulment in the face of sustained moments of surprise, when intersystemic communication appears, however internally, achieved. Urban 'reality' is a flexible, elusive tool of description that allows the movement from one urban construction to another, one system to another, one observer to another, one form to another, without marking any except its very own absence from the topology of the city. As such it signifies the reinstatement of the paradoxical, best observed through the utopian return to the system (the individual, internalised description that purports to thematise its known world, namely its systemic topology), and the consequent doing away with the need for a mitigating supra-system. It may also need to be further emphasised that the mediation of urban 'reality' resolutely indicates a theoretical tendency for systemic internalisation of all potential distinctions, rather than a reliance on existing societal markings. As explained below, the consequence of the above for the relation between the city and its environment is that the latter is no longer found outside the totality of communication as described by Luhmannian society, but simply and repeatedly outside the particularity of each self-undercutting representation of totality, namely each 'reality'. The incommunicability of the environment remains. What changes is the reason for the incommunicability: while in Luhmann the environment is incommunicable because it lies outside society, here it remains so because it lies outside the specific system of observation. The point of this shift is precisely the need to accommodate a system-specific ignorance in relation to both environmental law and the city, while exposing them to the horizon of potentialities.

III THE CITY NOT AS A SYSTEM

The city is not a social system, in the way Luhmann means it. It has its own distinct operations, it constructs its identity through self- and external reference, it produces its own meaning, it has an environment distinct to itself, but it is not a social system because it does not mark the 'communication' side of the autopoietic binarism communication/interaction. On the basis of this distinction, the city, in its abstraction, is defined here as an episode of intensification located on the boundary between communication and interaction. Communication and interaction lend their complexities to each other in a way that remain impenetrable to each other, and come together in what Luhmann has called a relation of *interpenetration*. I retain this term, in preference over the term *structural coupling* with which Luhmann later replaced interpenetration, precisely because of its spatial connotations, its facility to connect dissimilar forms of autopoiesis, and its ability to employ the other system's complexity as a resource rather than information.[67] Interpenetration is applied to the city as a spatio-temporal phenomenon that emerges in its systemic form as a terrain of boundary crossing. The city is defined, not in relation to its environment, but through a sustained self-reference that bifurcates (communication/interaction) for identity purposes, and in an organised attempt to exclude its environment, which lies in a relation of both continuum and rupture with the city. The environment is indeed excluded from the systemic organisation of the city, but emerges as threatening ecological risks from within.

More analytically, while communication is the operation of social systems (law, politics, religion, economy and so on), which necessitates the three elements of information, utterance and understanding, *perception* is its functional equivalent for conscious systems (humans) that are excluded from autopoietic society. Perception is a less exigent operation because it does not depend on being selected and communicated.[68] *Interaction* occurs when perception and its perception are being perceived, and this presupposes a reciprocal perception of perception amongst human beings present to one another. This transient way of communicating (indeed, the only way of communication) is what Luhmann calls *double contingency* and refers to the mutual internalisation of what each one thinks that the other has understood.[69] A boundary of an interaction system is defined along the notion of presence: 'they include everything that can be treated as present and are able, if need be, to decide who, among those who happen to be present, is to be treated as present and who not.'[70] Even if interaction systems produce communications, they will rarely feature as social communications. Thus, social communication and interaction remain distinct and confined to social and conscious systems respectively. However, social and interaction systems do come 'together' in a coupling of complexities that Luhmann, following Parsons, calls *interpenetration*. Interpenetration allows such associations, since it 'presupposes the capacity for connecting different kinds of autopoiesis – here, organic life,

---

67 The difference between interpenetration and structural coupling would have been destined to be a footnote in a historical account of Luhmann's terminology, had it not been for a very telling confusion on his behalf. See Luhmann, 1995, as to whether these concepts are worth differentiating or not (interpenetration as a precondition for or a variant of structural coupling). La Cour, 2006, has contributed greatly to the discussion with a differentiation between the two on the basis of the role of the environment: thus, in interpenetration, the environment is used as a resource, whereas in structural coupling, the environment is used as information. La Cour locates this differentiation on a second-order observation. I find that treating interpenetration as a *definitional* dependency, invisible to the system except as a form of invisibilised internal necessity, but on the level of operation, allows for the spaces of absence the present text is looking for.

68 Luhmann, 1995a:412.

69 Note that double contingency is not exclusive to conscious systems; Luhmann, 1995:392.

70 Luhmann, 1995a:412.

consciousness, and communication'.[71] Interpenetration's unilateral variation, penetration, occurs 'when a system makes its own complexity available for constructing another system'.[72] Accordingly, interpretation is the reciprocal availability of complexities,[73] which, however, remain peculiar to the original systems; there can be no fundamental common ground among systems.[74] Indeed, interpenetration is 'an intersystem relation between systems that are environments to each other'.[75] However, it has not been named inter-penetration lightly: 'the boundaries of one system can be included in the operational domain of the other . . . Every system that participates in inter-penetration realizes the other within itself as the other's difference between system and environment, without destroying its own system/environment difference.'[76] This mutual re-entry of each other's difference between system and environment guarantees that the two remain distinct.[77]

In Luhmannian autopoiesis, communication (social) and interaction (conscious) systems interpenetrate but can never combine to form one system.[78] Even in the ambits of the present analysis, where society is no longer and 'reality' informs interpenetration, the difference between communication and perception remains, the need for reciprocal availability of complexities in an unintelligible form is still relevant, and the mutual re-entry of the difference between autopoiesis and structure is always a prerequisite of interpenetration. But, despite the above, a formation does emerge.[79] I choose not to name the formation or indeed classify it. I will be calling it the urban system, for convenience's sake and because it possesses characteristics of an autopoietic system, but in full awareness that it would not be a social system in the Luhmannian sense. It is enough to conceptualise it without an epithet, as an episode that emerges when the two sides of the form communication/ interaction produce and consume each other without losing their autopoiesis.

71 Luhmann, 1995a:219.
72 Luhmann, 1995a:213, brackets omitted.
73 It is interesting how Luhmann, both with this passage and when explaining structural coupl-ing, seems to contradict himself by implying an ongoing autopoietic construction, which is at odds not only with his later writings but also with one of the main differences between him and Teubner, 1993. For an analysis of whether systems are or become autopoietic, see below, Chapter 6.
74 Luhmann, 1995a:35.
75 Luhmann, 1995a:213.
76 Luhmann, 1995a:217.
77 Consequently, Luhmann, 1994c:380, defines socialisation as 'the process that, by inter-penetration, forms the psychic system and the bodily behavior of human beings that it controls'. See also Vanderstraeten, 2000.
78 'Communication operates with an unspecific reference to the participating state of mind; it is especially unspecific as to perception. It cannot copy states of mind, cannot imitate them, cannot represent them' (Luhmann, 1994c:381).
79 Although quite different to *symbiosis* as suggested by Maturana and Varela, 1992.

Episode here should be understood as an event that interrupts while continuing a process.[80] An episode is instrumental to the continuation of the reciprocal autopoiesis but also extraneous to either communication or interaction. Here, episodes should be understood as the temporal emergences of interpenetration, where the one side of the form 'reveals' (but always internalised as own) itself to the other as its boundary. Interpenetration is a fragmented series of episodes of this sort, blindly received by each side of the form communication/interaction, and without any promise for continuation. The way episodes reveal themselves is through an intensification on the other side: of the social if conscious and of the conscious if social. This is not observable (only) on the second-order level, but it is an ontological construction (always intentional) of a catenation of unconnected but persistent episodes clustered together on account of their historicity and spatiality. In that sense, interpenetration reveals itself as the temporal and spatial binding (that is the preservation of meaning over time and across space), of the parallel revelation of episodes between communication and interaction, creating thus the astonishment of 'reality'.

The city, then, is not much more than an unconnected exchange of episodes between the two sides of the urban form communication/interaction. The surprise at the resulting continuity despite fragmentation has been matured into an expectation of 'reality', and as such thematises the connection between an observer and the urban system. In other words, urban 'reality' is located on the distance between the observer and the city. This space is where urban autopoiesis is to be located – neither in the institutions of the city nor in its neighbourhoods. Autopoietic space – old and new, present and potential, fragmented and continuous – 'is curved and closed in the sense that it is entirely specified by itself'[81] and its only operation is to emplace the autopoietic system in its environment. Urban autopoietic space is where the temporality of the urban episodes materialises. The space of the city binds time in that it includes, in potentiality and as part of the system's structural predetermination, its evolution.

Once again, I turn to Calvino for an example, this time his description of Olinda:

> If, when in Olinda, you go out with a torch in hand and look carefully around, you might find somewhere a point, not bigger than the eye of a needle, in which, when looked at through a magnifying glass, the roofs, the aerials, the skylights, the gardens, the ponds, the street signs, the corner shops on the squares, the racecourse can be seen. This point does

---

80 Luhmann, 1995a:407, talks about interactions as *episodes* of societal process, but not the other way.
81 Maturana and Varela, 1972:92.

not just stand there: a year later it can be found big as half a lemon, then as an edible mushroom, then as a soup bowl. And there it is, a city of natural size enclosed into the previous city: a new city that grows by pushing the other city outwards . . . In Olinda . . . the old wall of the city is expanding, dragging along the old quarters, which grow on a broader horizon at the confines of the city, while at the same time maintaining their proportions . . . [A]nd in the centre of the innermost circle – however difficult to discern – the next Olinda and those that are going to grow after it are sprouting up.[82]

Olinda is a city like any other. Cities grow from inside, impregnate themselves, regurgitate the same urban structures, push their boundaries without altering their autopoiesis.[83] Cities themselves produce their boundaries along whatever horizons are incised within their materialities. Olinda produces itself in a typical autopoietic fashion, where the city changes but the urban system does not. Biological autopoiesis reserves a special role for autopoietic space.[84] The latter coincides entirely with the processes of autopoietic production of the system, that is the operations peculiar to the system, as produced by their consecutive 'superimposition'. The topology of the system is coterminous with its operations and their production. There is no restriction to the variations of the topology of the system (the circles in Olinda can be inner or outer), as long as its autopoiesis is maintained throughout the variation (as long as, that is, the proportions of Olinda are preserved despite its growth).[85] The spatio-temporal fluctuation of Olinda does not change the identity of the city: the persistent production of its operations (streets, turrets, wanderers . . .) holds the city together. Production, of course, is only one side of the topology, the other being consumption of the same operations. While processes of production have been used extensively in relation to space,[86] a complementary understanding in the manner of consumption has only recently emerged as part of a post-Marxist, postmodern geography.[87] Autopoietic circularity between being and becoming demands a diffusion of emphasis and a parallel marking of consumption when it comes to

---

82 Calvino, 1993:130–31.
83 The city's 'thermodynamic openness', to the point of absolute dependence on its ever-expanding 'global pastures' (Girardet, 1992; Haughton and Hunter, 1994; Folke and Kaberger, 1992; Prigogine and Stengers, 1984; Kauffman, 1995) does not preclude autopoiesis. According to Maturana and Varela, 1972, the autopoietic unit is thermodynamically open; only its operations are closed.
84 Maturana and Varela, 1972:90*ff*, employ the distinction relations/components, whose circularity defines the topology and identity of the system.
85 Maturana and Varela, 1972:93.
86 See Lefebvre, 1991; De Certeau, 1984; Harvey, 1996b.
87 Baudrillard, 1998, employed this inversion in his genealogy of consumption, in an attempt to describe the motivation behind a globalised society.

autopoietic processes. Their relation is not simply one of circular causality (consume in order to produce/produce in order to consume), but one of difference of difference: at any one point, the autopoietic system produces and consumes meaning, constitutive processes, topology, and history: in short, an autopoietic system simultaneously produces and consumes itself. This attributional way of perceiving circularity lends an illusion of coherence to the episodic nature of the urban system.

Whatever the illusion and however utopian 'reality' may be, the city remains a series of episodic intensifications, immured on each side of the form. Luhmann accepts that an intensification of communication is possible within interaction systems, but such intensification is only understood by communication, and then only as an internal operation.[88] This means that intensification of either is only understood on the other side of the form. Following this, one can theorise intensification as the internally constructed escalation of spatio-temporal production and consumption processes, which strengthens the self-reference of the system (the one that becomes intensified – not the one that triggers the intensification) by engaging it in a spiralling resemiologisation of its topology that does not result in a loss of identity.[89] Intensification is felt as a reverberation on the opposite side of the boundary between communication/interaction, and is intricately linked to the topology of the system. Accordingly, what binds these systems together is the mutual triggering of their spatio-temporal intensification, a sort of desire for hallucinatory substances where addiction is only confined to the desire. The city is a series of such episodes of intensification, or, as Deleuze and Guattari would put it, instances of 'reterritorialization'. For Deleuze and Guattari, territories are emergences of combinations of various 'milieus' (namely, the material field on which assemblages of populations are formed – a qualitatively different equivalent for *meaning*) that provide the feeling of being 'at home';[90] and 'reterritorialization' is precisely a new combination of milieus that, following 'deterritorialization', forms a new territory. In this process, with its external ('de-') and internal ('re-') reference, the binding element is not the species that moves, but the territory on which the assemblage of species is formed.[91] In urban autopoiesis, the territory changes according to the way meaning is internalised by communication/interaction and employed in the production/consumption of

---

88 Luhmann, 1995a:414.
89 Intensity in thermodynamics is the state at which a system is pushed beyond its normal thresholds, where average homeostasis could operate, and into a zone of crisis, where new patterns and thresholds are employed.
90 Deleuze and Guattari, 1987:311.
91 In fact, the relation between the 'de-' and 're-' of territorialization, namely, the 'line of flight', Deleuze and Guattari, 1987:55. This is a simplification, since it should refer to the interaction between space ('striated' and 'smooth') and the bodies or assemblages that have territorialized it. The discussion on corporeality and autopoiesis follows in this section.

topology – but the kind of meaning employed (or excluded) is determined by the topology, in a circularity whose unacknowledged purpose is to be 'at home', to adjust novelties to existing patterns of autopoietic evolution while ensuring that the intensifying effects of interpenetration remain.

On the social side of the urban form, the usual operations of the various social systems are to be found, with the following proviso: social systems are not drawn together by an overarching idea of emergence, not even on a second-order observation level. Their differentiation from interaction systems is based on the difference between communication and perception/interaction, not on a defining line of society. While this may seem not too different to the Luhmannian idea of a determinable, contingent society, the difference is indeed important: the system (each system) is immediately exposed to the potential intensification coming from interaction; the boundaries between communication and interaction cannot be discerned with certainty, especially at the points of intensification; and the series of episodes that constitute the urban system can be classified as a systemic formation, without the need to be thought of as belonging either to society or outside. For the time being, therefore, the concept of communication can be kept in the way Luhmann describes it – its resemiologisation will become relevant in the next chapter, when the distinction communication/interaction will be problematised by other, non-meaningful operations.

On the other side of the form, interaction should be defined in a more complex manner than Luhmann's interaction systems, for the reason that the urban form is characterised by a meaningful materiality of the kind that Luhmannian autopoiesis would not accept. The problem is located in the fact that Luhmann differentiates between conscious and bodily systems, and attributes very little importance to the latter, except in their possibility of being 'steered' by the former.[92] Even then, 'the difference between corporeality and noncorporeality has (at least for our present societal system) no social relevance'.[93] This is rather disappointing for a theory that insists on being able (and to some extent managing) to break free from subject/object binarisms, and instead falls headlong into the most widely criticised Cartesian division between mind and body.[94] Admittedly, the mind is not Luhmann's prime preoccupation. However, he has defined the mind and has repeatedly dealt with its representations and its connection to communication – so, opportunities to deal with its materiality have frequently occurred.[95] Indeed,

---

92 Luhmann, 1995a.
93 Luhmann, 1995a:246.
94 e.g., Plumwood, 1993.
95 Luhmann, 1994c, 2000a. Consideration of corporeality would not endanger the basic autopoietic distinction between society and human beings, since it would have stayed on the environmental side of society. The body question is clearly a matter of uneasiness for Luhmann, and this is nowhere more obvious than in his treatise on love, 1998b.

in *Ecological Communication* he writes: 'ecological relevance for society is mediated by its relevance for the human body, possibly heightened by perceptions and anticipations, that is, by psychic [conscious] mechanisms. In thinking about destruction, it makes no sense to think of people and society separately.'[96] But the body did not appear in that analysis either. A possible reason for this indifference is the body's situatedness in space – and arguably only for space does Luhmann reserve a greater indifference than for the body.[97] However, the refusal to deal with space is very difficult to justify in view of the numerous pages he has dedicated to questions of temporality, considering the fact that nowadays, any insistence on considering the two separately cannot but be regarded as antiquated.[98] Since both space and body are relevant in the present definition of the city, attempts will be made to reconcile the theoretical departures with at least some of the autopoietic axioms, but not to the point of endangering the solidity of this volatile theoretical construction which is the city.

Phenomenology, also because of its elective affinity with autopoiesis, can be a source of succour.[99] Husserl has linked body, space and perception through the concept of *kinaesthesia*, which emphasises subjective movement in space as the mode in which perception of our body, objective things and even reality takes place. In this vein, it has been suggested that only through corporeal movement does the subject become subject,[100] and that indeed movement, more than perception, should be considered the paradigmatic experience of mental intentionality, namely connectedness with the world.[101] Following Husserl, Merleau-Ponty accentuates the relevance of movement and describes the mind as operational only through the body and in relation to the objects around it. '. . . [T]he subject penetrates into the object by perception, assimilating its structure into his substance, and through this body the object directly regulates his movements. This subject-object dialogue, this drawing together, by the subject of the meaning diffused through the object, and, by the object of the subject's intentions, arranges round the subject a world which speaks to him of himself, and gives his own thoughts their place

---

96 Luhmann, 1998a:83.
97 This is not a question of cultural, national, ethnic space, etc. which is included (and relativised) in the concept of *world society*, Luhmann, 1990c, but space as a correlate of the system's organisation, in the same manner as time. Passing references to space, but mostly to comment on what Luhmann sees as the limited analytical relevance of space, can be found in Luhmann, 1997a and 1998a. See also Balke, 2002, for his spatialised version of autopoietic exclusion.
98 See indicatively, May and Thrift, 2001; Soja, 1989; Harvey, 1996b; Grosz, 1995; Massey, 1992.
99 More contemporary approaches, such as Haraway, 2004; Braidotti, 2002; Strathern, 1992, are also relevant – see below.
100 Petit, 1999.
101 Pachoud, 1999.

in the world.'[102] It is interesting to see how, despite a general attempt at diffusion of direction, the centre of this dialogue remains the subject. Husserl was less indirect: in his famous *epoche*, namely the passage from the Natural (i.e., everyday, first-order) to the Transcendental (i.e., philosophical, second-order) Attitude, the one thing that resists bracketing is the ego, because bracketing presupposes the ego. Reminiscent of the autopoietic act of distinction, which is always non-transparent to the observer who performs the distinction,[103] the unbracketable ego – or 'transcendental consciousness'[104] – is the central point of phenomenology, upgrading phenomenology to what Husserl somewhat inelegantly calls *egology*. 'The ego is the center from which all the different acts flow and toward which all the different affects stream.'[105]

Bell, a commentator of Husserl, in a manner reminiscent of Merleau-Ponty,[106] exalts the body, as it stands in the realm of reduction, to the only comparator against the claustrophobia of the Transcendental Attitude: 'now my own living body is the sole absolute point of reference, the unique "geometrical centre", of egocentric or orientated space.'[107] The body, a centre of the intentional activities correlative to the intentional space, egocentrically thematises the space in which it moves, so much so that spatiality becomes a quality of the body itself: '*der Leib als Orientierungszentrum.*'[108] And 'regardless of how much objects may change their spatial positions, and regardless, too, of how much I might change mine, a firm zero of orientation persists, so to speak, as an absolute Here'.[109] The egological body (the 'I') is the absolute spatial centre, the transcendental here,[110] just as the system is the absolute centre of its 'reality', thematising itself, its operations and its environment with its individualised, egological meaning.

Regardless of the emphasis shift from subjectivity to social systems, the connection between phenomenological and autopoietic solipsism is evident.[111]

---

102 Merleau-Ponty, 1962:132, reiterates the Husserlian intentionality of consciousness, but with limited references to the language of consciousness in his attempt to avoid the mind/body dualism, of which Husserl himself, despite his efforts, has been found guilty (e.g. Macann, 1993), since he replaces the Cartesian mind with phenomenological consciousness while retaining the body-mind binarism.

103 Luhmann, 1998a, after Spencer Brown, 1969.

104 Husserl, 1973a.

105 Kockelmans, 1994:121.

106 See Merleau-Ponty, 1962, where the body has what Merleau-Ponty calls 'motor intentionality', which is comparable to Husserl's 'kinaesthesia' as it appears in Husserl, 1983.

107 Bell, 1990:221.

108 'The living body as the centre of orientation', as quoted in Bell, 1990:210, n. 16, from Husserl, 1954.

109 Bell, 1990:210.

110 Husserl, 1973a.

111 Despite genuine efforts of both disciplines to avoid subjectivist closure by taking recourse to concepts such as intentionality and intersubjectivity for the former, and concepts such as second-order observation, structural coupling, double contingency and so on for the latter,

What is of interest here, however, is the way phenomenological solipsism manifests itself in the way the body emplaces itself in space. This does not entail a reinstatement of subjectivity, a novel take on consciousness, a new fiat of humanism or indeed a sneaky helping of free choice. If intentionality is understood as the delimitation of difference in a non-prioritisable identity formation, then intentionality expresses, rather than a revival of humanism, its very limits.[112] In an autopoietic city, the body and its intentionality are not features of human agency. On the contrary, they express the impossibility of isolating the human as object of observation, let alone as agency. Through intentionality, the body thematises a space in the Husserlian egological way, while at the same time being thematised by this space. Intentionality, as the parallelism of incommensurate differences, allows one to inform the second part of the Husserlian proposition with developments in spatiality and corporeality that take into consideration the processes of intensification as originating in social systems. These developments, largely originating in feminist preoccupations with spatiality,[113] are indispensable to the autopoietic definition of the episodic urban form, because of their emphasising, in their own way, the role of social structures in identity formation. To these developments the text turns briefly, in order to describe the interactional side of the urban form and connect it to its intensification.

While identity is not part of the present quest,[114] it is easy to see how it relates to interaction systems. Through contemporary attempts at identity descriptions, the interpenetration between the social and the interactional is confirmed. Thus, Henri Lefebvre's understanding of the body's emplacement in the urban space: 'each living body is space and has space: it produces itself in space and it also produces that space.'[115] Lefebvre's seminal work on rhythm-analysis has revolutionised the way corporeality and spatiality are combined in order to eavesdrop on what he calls the city's real meaning, as situated beyond narrow objectivity and well into the implosive domain of senses as sensual and sensible. This was the beginning of sensuous geography,[116] that

---

the avoidance is not altogether that successful, especially in view of retaining concepts such as egology and philosophical attitude for the former, and binary coding and 'unmarkable' environment for the latter.

112 *Contra* Nelson, 1999.
113 E.g., Massey, 1994; Villanueva Gardner, 1999.
114 But see Philippopoulos-Mihalopoulos, 2006.
115 Lefebvre, 1991:170; see also Latour, 2004b.
116 Rodaway, 1994:5 'The sense(s) is(are) both a reaching out to the world as a source of information and an understanding of that world so gathered.' The term 'sense' is used here both as sense mode (olfactory, auditory, haptic, visual) and as meaning ('to make sense'). See Feld, 1996:91, '[a]s place is sensed, senses are placed; as places make sense, senses make place'; Casey, 1996:24, 'just as there are no places without the bodies that sustain and vivify them, so there are no lived bodies without the places they inhabit and traverse'; and Tuan, 1977.

emplaced the body in its geography, a manifestation of which can be seen in Richard Sennett's urge for the body to reappropriate the city and allow itself to be mapped while mapping.[117] The circularity between body and space has been emphatically observed by Judith Butler in her analysis of performativity, as well in her later citationality thesis,[118] where the practice of ritualistic iteration provides for a gendered identity that can be best seen in 'the mundane way in which bodily gestures, movements, and styles of various kinds constitute the illusion of an abiding gendered self'.[119] Butler's performativity has opened up a whole discussion on the importance of emplacement, echoes of which one finds in the importance of movement for the phenomenological being. Thus, the Rio Carnival performances constitute space, as well as being constituted by it, reinstating process over agency;[120] dancing in the city, on the other hand, through the very repetition that according to Butler obliterates agency, seems to allow femininity to acquire force and motility, and from powerless construction becomes powerful subject.[121] In Briginshaw's view, the bodies conflate with the city, conferring to it the organicity of place, while the city envelops the bodies setting up for them an angular but able-to-be invaded space: 'the boundaries and zones of the performers' bodies are partly constituted through their linkages with the surfaces and planes of the city . . . Through this process of mutual definition the dancing bodies become subjects and the spaces of the city settings become places.'[122]

Ways out of the impasse (by delving deeper into it) are offered by Gillian Rose's unfolding of the imaginary next to real space as constitutive of a spatial paradox which needs to be imagined in order to sustain the fluidity between the two (and resist masculine separation).[123] Thus, as Rose elaborated in her later texts, space becomes fantasised and corporealised through a performativity that necessarily reveals 'gaps' and fissures, in a circularity of inside and outside, where the margins seem a better place to be.[124] This movement between margins and centre is also observed by bell hooks in her understanding of politics as location and articulation, namely the spatiality and alternative voicing of identity.[125] In hooks's account, corporeality is once again interdigitated with a racialised form centre/margin, where the latter remains marginal as the only space of resistance. But margin often appears not outside but inside, incised on the horizons of the city, the lines whose

117 Sennett, 1995.
118 Butler, 1993.
119 Butler, 1990:140.
120 Lewis, 1996.
121 Briginshaw, 2000:118; See also Thrift, 1996.
122 Briginshaw, 2000:110.
123 Rose, 1993.
124 Rose, 1996 and 1999.
125 hooks, 1990.

mutual crossing constitutes the urban experience.[126] Likewise, for De Certeau, walking in the city is resisting the structures of control and discipline: '[t]he Concept-city is decaying . . . one can analyze the microbe-like, singular and plural practices which an urbanistic system was supposed to administer or suppress, but which have outlived its decay.'[127] De Certeau subscribes to an idea of conflictual co-determination of the body and the city: the Marxist body of the urban *flâneur* as an atomic piece of resistance against the organised urban capitalism. This resistance is deemed successful: 'spatial practices in fact secretly structure the determining conditions of social life.'[128] The practices include narrating, name giving to streets, train travel, all of which carry out the same act of resistance against (but always as part of) the topological production and consumption of the urban system.

In a not dissimilar vein, Elisabeth Grosz's rendition of the body/city-space relation, resistance takes place precisely in the way the one produces and consumes the other, in the way 'they are mutually defining'.[129] Instead of a causal or representational mode of connection, Grosz suggests 'assemblages or collections of parts, capable of crossing the thresholds between substances to form linkages, machines, provisional and often temporary sub- or micro-groupings . . . Their interrelations involve a fundamentally disunified series of systems, a series of disparate flows, energies, events, entities, bringing together or drawing apart their more or less temporary alignments.'[130] In her reading of Deleuze in relation to architecture, Grosz describes the body/city-space as a series of eventful constellations, involving a constant crossover of meaning from social structures to interaction and vice versa, which resists typification and ends up as an inalienable but defining moment of the difference between them.

All the above, and this was only a representative slice of the booming theoretical energy in this domain,[131] are certainly (but not simply) ways in which conscious systems, through their very corporeality and spatiality, interact. The requisite of presence in interaction is mediated by space and body: thus, autopoietic interaction is now taking into account the generation of perception in the mutuality of corporeality and spatiality, thus revisiting its existing prioritisation of mind and time. As contextualised in urban space, interaction triggers moments of intensification of the social side of the form. Likewise, law, politics, economics, science and so on, are thrown into a

---

126 Weiss, 2005.
127 De Certeau, 1984:95–96.
128 De Certeau, 1984:96.
129 Grosz, E., 1995:108; also Grosz, 2001.
130 Grosz, E., 1995:108.
131 The law has been involved here too: see Manderson, 2005; Johnson, 2005; Valverde, 2005, and the entire volume 9 of *Law Text Culture*, dedicated to Legal Spaces; Philippopoulos-Mihalopoulos, 2007, and the introduction therein; and Taylor, 2006.

(self-referential) deciphering of the difference between their external and self-reference, which reaches levels of intensification in the urban episodes. The conscious penetrates the social, elevating presence into a systemic topology of mutuality between body and space. The social penetrates the interactional, igniting frictions of propinquity, illusions of concrescence, fractures of alienation, desires for intimacy. Interpenetration is always blind, without origin, causality or purpose, consistently self-produced. The city concentrates the difference between social and interactional in one fragmented series of interpenetrational episodes. The city constructs itself on the boundary between communication/interaction, 'a city balancing on the tip of a pyramid'[132] perilously moving without moving from intensification to intensification, expanding or decaying in an evolutionary spiral of self-referential ignorance, whose source is nowhere to be acknowledged, except perhaps on the very difference between the sides of the urban code.

Is, then, the city a system? Yes, to the extent that it produces and consumes its own topology; that its closure is preserved by a selectivity of meaning peculiar to its coding; that its code is a spatio-temporal contextualisation of communication/interaction, which guarantees its self-reference while eluding its self-description; that it is produced in the intentional distance between its signs (operations) and its observer; that its unity is not dependent on anything external but instead on its immanent consumption and production as apprehended by an observer whose 'reality' is conditioned by surprise; that its ontology is confounded with its epistemology; that, finally, its reality is the marked difference between the totality of descriptions and the absence of descriptions. But the same can be answered in the negative. Thus, the city is not a system to the extent that it is neither social nor interactional; that its operations are always the unmarked, other side of intensification – whatever triggers it, be it communication or interaction; that its code is a spatio-corporal boundary between communication/interaction, selectively apprehending only the episodes of intensification that construct the urban 'reality', which, at the same time, defines and marks these episodes as pragmatic, despite their utopianism.

The city is a *durée*, a long movement in time and space, a Bergsonian flux of becoming whose points and stages are not discernible, whose mobility denies its very spatiality and temporality, whose existence is a paradox in view of the incommunicability of the points; yet it is here, a Notre Dame reconstructed, Paris assembled,[133] despite its mutually undercutting horizons inscribed in the urban folds. A terrain of observation as shifting and fractured as that is preferable to a supra-system that delineates the cognitive horizons of its sub-systems and domesticates a slice of the environment

---

132 Calvino, 1994:73.
133 Bergson, 1955:32.

according to its limits of uncertainty. Here, boundaries are always the other side of what they seem, mutuality is always undercutting, and paradox remains folded yet operable.

IV THE CITY IN ITS ENVIRONMENT

On such a terrain, one looks for the environment. But the terrain is not just unstable; it is also impenetrable. The code communication/interaction is seemingly all-inclusive, in its extension it becomes a form, in its marking a self-indulging game of mirrors. Where is the environment of a form? What is outside the boundary of a boundary? How large a leap must it be in order to circumvent both sides of the form and reach beyond? The environment of a form is nowhere to be found, for a form needs to be marked in order to become operable. From the form, a system and its environment will emerge. Next to the form, nothing. But then, from the nothing emerges the first distinction. To put it plainly, the environment of the form is whatever severs the form, tips it off-balance, renders it observable. It is an environment always to come, already implied in the form, and perennially absent. The city is busy marking itself this and that side of the boundary, each one a difference of system and environment for the other, impervious to whatever happens outside, happily internalising its material environment, devouring its global pastures, constructing itself and nothing else. The city has forgotten its outside, and nowhere is the environment more absent than in the city.

Regardless, the city is constantly tipped off-balance. Whether it is because of an actual outside, a potential outside, or a genuine inside (but of the other side), it is indifferent. The operation is always internal. The city is resilient but exposed – it has no society around it to protect it, no means of closing itself against things that it does not understand. In its bourgeois arrogance, the city is like a rose on another planet: attractive but defenceless.[134] And wholly incapable of dealing with the greatest threat of all: itself. Its very acceptability, volatility and fragmentation instigate its risks and opportunities. This is precisely the environment of the city: the potential continuum between the city and its environment, its bodies and other bodies, within and out, its space and its pastures. The environment of the city is not *just* nature (what is this?), ecological catastrophes (where do they come from if not from within?), bodies that disrupt the circularity between space and corporeality, monsters and ignorance; *but the system's very possibility of including all these as part of its systemic topology*. In other words, the environment for the city is the immanent, irreversible, unquenchable continuum between the inside, where the pleasantries of the form are to be found, and its outside, namely everything

---

134 On the little prince's planet: de Saint-Exupéry, 1946.

that could potentially sever the form, observe it from the outside, and from its vantage point cut the *durée*, totalise its coherence, and institute margins inside the form. There is no goodie and baddie here – just an irrepressible desire for continuum.

An environment of continuum makes the division culture/nature redundant, the need to define ecological processes (human or natural?) ancillary, and the conceptualisation of the human superfluous. This emphatically does not mean that there *is* a continuum. It means that anything that is potentially understood as continuum between the city and its environment is in the environment of the city, expelled from the urban form but always part of it. The emphasis is on the contingent, the meaning before/after its actualisation, an autopoiesis sat on its head: the emphasis shifts to the environment, but in its harrowing absence. And this is where the flipside appears, the other side always already implied in the first side, the other side which is not other: *continuum is a phenomenon of a continuous rupture whose serialisation and superimposition appears as connection.* The understanding of an environment of continuum/rupture cannot rely on ecological theories that urge for the recognition of affinity between human and natural,[135] on discourses of nature exploitation,[136] on a revived ecological humanism that believes itself to be instrumental for an enlightened anthropocentrism or inclusive ecocentrism,[137] or on pragmatic combinations of economics and environmental protection, such as sustainable development. All these are simultaneously supposed solutions and insurmountable problems, because they all start from a concentric concept of relationality, a central structure that needs to be moved, a margin that needs to be introduced. But the environment of the city does not – cannot – do any of these. On account of its continuum/rupture with the city, the environment *is* the city, in a cupola that continues and fractures itself as soon as uttered. Giorgio Agamben's description of *polis* and *physis* is resonant of precisely such a form: 'the state of nature is not a real epoch chronologically prior to the foundation of the City but a principle internal to the City, which appears at the moment the City is considered *tanquam dissoluta*, "as if it were dissolved" (in this sense, therefore, the state of nature is something like a state of exception).'[138] Nature is the exception that has become the rule, the non-connection that connects. The city is a process, wholly ecological, wholly human, wholly cultural, wholly natural, wholly perforated. Its environment is a striated space echoing of smoothness,[139]

135 Ecological processes are seen to include human processes, e.g., Wiener, 1995; Orts, 1994; for an interesting declaration of affinity, Game, 2001; for more specifically urban processes, see Harvey, 1996c, but also Clark, 2000, for a complexity-informed description of cities.
136 Plumwood, 1993.
137 See esp. Wilkinson in Holder and McGillivray, 1999.
138 Agamben, 1998:105.
139 Deleuze and Guattari, 1987.

folded up to contain the city's multiple horizons, organised and ordered in order to maintain the beast at bay, chaotic and primeval in order to provide for the city's hinterlands. The environment of the city is all that the city is, making the city inescapable and the environment absent.

'A big chunk of nature falls in the middle of the city.'[140] Michel Serres relates the trial of Anaxagoras, who was taken to court for suggesting that the sun was a hot stone. The city could not accept this, and Anaxagoras was due to be ostracised. But the chunk has fallen in the city, and the city has been tipped off-balance, severing its limbs and kicking its citizens out of its walls. Serres's famous natural contract would have been relevant here, had it contained a forfeit clause. But in the *polis* there can be no symbiosis and reciprocity – just a frightening and exhilarating continuum that threatens to upset a fragile catenation of episodes and reveal all its ghosts – for they are already in the city. The city accepts within its precarious walls all its environmental monsters: it remains the luminous *accueil* and purgatory of the 'subaltern'.[141] The city is as 'guilty' as the law for wanting to carry on its surviving autopoietic act: just as the law, the city constructs an environment, domesticates it, awards to it an external reference, and carries on busying itself with its own operations. It cannot do otherwise. So the city climbs on the stone, puts a flag on it and builds a cinema showing films of exotic locations. The city converts its absent environment into a domesticated presence that signals absence. This has an appeasing effect for the city, but keeps on circumventing the environment.

The environment as continuum/rupture has been described frequently. It is invariably characterised by a certain nonplussed embarrassment as to its limits, which is sometimes offered as an anathema, and at others as a blessing. One of the most classic examples comes from Henri Lefebvre: 'The urbanites located themselves by reference to the peasants, but in terms of a distanciation from them.'[142] This is perhaps the simplest form of continuum (of reference)/rupture (of class/location/habits), which has been substantially enriched by the ecological paradigm. Paul Virilio, for example, foresees a total absence of nature in his 'teletopia' (the disembodied reality of a telematic society), both in the sense of natural habitat and the more abstract notion of nature as affinity between humans and the rest of the planet.[143] But he shifts to a high note of sorts when he writes that this total schism between human and natural as brought about by an all-determining science, will make humans realise their inability to live without nature and in separation from the cosmos, and consequently reject a technology that denies the materiality

140 Serres, 1995:71.
141 Spivak, 1988.
142 Lefebvre, 1991:268.
143 Virilio, 1993; Baudrillard 1998; see also Rolston III, 1988; Gardiner, 1990.

of the body and its environment. On the other side of catastrophe one finds Donna Haraway, who declares that 'the boundary between human and animal is thoroughly breached' and whose cyborgs, oncomice and coyotes are posthumanist versions of socialist and feminist personas that transcend the natural/cultural, organic/mechanical, physical/non-physical and so on divides.[144] In a not dissimilar vein, although from a different point of view, Latour talks about 'hybrid networks' between social, informational and ecological systems,[145] and the 'pluriverse' consisting of collectivities of humans and non-humans that redefine democracy as something that can be found either side of the boundary.[146] In her study on the urban domestication of monsters, Ruddick puts it eloquently: 'for all our hopefulness about the progressive potential of cities, we are on the verge of a dangerous fusion which threatens to combine, in our worst nightmares, the stories of monsters as excentric and monster as intrinsic.'[147] This fusion is the way both into and out of the nightmare, a rite of passage which can only be deparadoxified through an employment of excentralised absence. Finally, Agamben locates the *Homo Sacer*, the bandit and the werewolf right there, on the threshold between continuum and rupture: 'what had to remain in the collective unconscious as a monstrous hybrid of human and animal, divided between the forest and the city – the werewolf – is, therefore, in its origin the figure of the man who has been banned from the city . . . a threshold of indistinction and of passage . . . who dwells paradoxically within both while belonging to neither.'[148]

These accounts have in common an excentralisation of habitually centralised concepts – human subjectivity, agency, technology, science, ecology, and so on. Such endeavours can only be sustained if they take place within a wider context of excentralisation of the centre itself, which occurs via an emphasis on the 'crossing', the continuous exteriorisation (or internalisation) of meaning, the 'polluting' of the other side. The autopoietic way of achieving this is through the paradox of disrupted continuum between system and environment. The case of corporeality, as both inside the urban system (as defined here) and outside any social system (as defined by Luhmann), is an example of how the continuum between the two sides (inter- and extrasystemic) of a thing that does not exist ('nature', city, environment) is brought forth. The schizophrenic definition of ecological processes as both natural and human (isn't human natural?), the extension of environmental protection claims to human-made environment, the impossibility of encountering a totally natural (that is, non-human) environment: all these are aspects of the environment as continuum/rupture either side of the boundary.

144 Haraway, 2004; see also Braidotti, 2006.
145 Latour, 1993; for a contextualisation of hybridity, see Grabham, 2006.
146 Latour, 2004a.
147 Ruddick, 2004:36.
148 Agamben, 1998:105.

The need to see the environment as continuum/rupture is enhanced by the illusion of systemic all-inclusiveness: if the system feels that it includes everything that can be included at any one time, there can be no novelty in the system.[149] But where is ignorance to be located in that case? Where is the passage? Virilio's teletopia, for example, is an ecologically menacing version of the cybercity discussion: 'constructed spaces will increasingly be seen as electronically-serviced sites where bits meet the body – where digital information is translated into visual, auditory, tactile or otherwise sensorily perceptible form, and vice versa'.[150] Indeed, cybercities have been hailed rather apocalyptically: '[n]ow, a simulacrum of the city is growing in cyberspace. This virtual city is ramifying through the real city and in the process reproducing it.'[151] It is very doubtful, though, whether these elements can be considered 'new'. Cybercities anagrammatise geographical cities: whereas in the latter, the human being articulates their biological and social duality,[152] in the former, biology is incorporated in technology and produces a habitus in which humans imagine their bodies as defined by the technological reproduction of sensory perceptions.[153] One's mind invents one's body and this invented body is communicated immaterially through technological channels to reach other immaterial bodies who will imagine and project the received body according to their very own specifications. There is no novelty, just a resemiologisation of existing operations between the city as geography and the city as a catenation of events, 'asynchronous in its operation, and inhabited by disembodied and fragmented subjects who exist as collections of aliases and agents'.[154]

A similar example can be found in the case of the global city, whose continuum seems to embrace the globe – which includes and feeds into the theory of globalisation itself – while its rupturing gives rise to localisations and reinstatements of community nostalgia and illusions of local control.[155] A global city swallows the world in a telematic megalomania, and allows no margin for rupture within its self-description.[156] Globalisation theory, at its most assertive, not only describes a future, but constructs a present, and a lukewarm present for that matter: globalisation attempts simply to shift things around by naming them differently. By postulating that control

---

149 Luhmann, 1995a, 1993a.
150 Mitchell, 1994:16.
151 Pickering, 1999:181. In the same vein, see Downey and McGuigan, 1999; Graham and Marvin, 1996; Plant, 1997; Virilio, 1993; Graham, 1999.
152 A point dealt with from several perspectives: see Caldwell, 1990; Derr, 1975; Gardiner 1990.
153 Chatterjee, 2006.
154 Mitchell, 1995:24.
155 Bauman, 1998.
156 Connelly and Smith, 1999:195; see also Scholte, 1996; Beck, 1996; Sassen, 1991, for instances of total continuum.

changes its level of operations from the local to the global,[157] globalisation relies on the power paradigm and perpetuates hierarchies and illusions of control.[158] Of course, the construction of future as present is more often than not effective, as it is shown by an example from *'lex mercatoria'*, which is another domain in which globalisation seems to construct itself quite effectively. *Lex mercatoria* is the nascent, but 'achieved' set of principles of international and transnational merchant law that are applied by decision makers in the same way as any other recognised *lex fori*.[159] When Carbonneau admits that it is unclear whether the main treaty in the *lex mercatoria* area, that of the 1958 New York Arbitration Convention, simply recognises *a posteriori* a generalised international practice, or introduces new elements, he acknowledges that the difference between potential future and constructed present is insignificant when it comes to something so self-legitimising as *lex mercatoria* or globalisation in general.[160] Thus, globalisation theory legitimises – if not constructs – globalisation, just as globalisation legitimises the existence of globalisation theory. Its multifarious articulations have managed to produce a self-referring discourse that perpetuates its self-importance: 'The question of globalization succeeds in conveying its own importance to the extent to which it imputes importance to those who are involved in discourses about globalization.'[161] In essence, globalisation adopts the well-digested, non-demanding paradigm of hierarchical control, and expands it to cover literally the globe, thus converting it into its own central tenet.[162] Globalisation does not offer a new description of the world; instead, it constructs itself as the spatially and temporally enhanced hierarchical paradigm which imposes itself onto the present, thereby attempting to alter the present according to its omens: a self-fulfilling prophesy par excellence.[163]

In discussions on both cyber and global cities, novelty is covered up as non-novelty, and absence is swallowed up by the name-giving system. Non-novelty means that there is a blind mirroring of the system into its environment, a sort of tautological continuum between system and environment, so that nothing that is not in the system is not *of* the system – hence, nothing escapes the legitimation of naming. But this non-novelty, which is indeed an exercise

157 Beauregard, 1995.
158 Low, 1996, and generally Cox, 1997.
159 Highet, 1990:101.
160 Carbonneau, 1990.
161 Schütz, 1996:271.
162 Fitzpatrick, 2001.
163 Hirst and Thompson, 1996, Chapters 2 and 3, posit that globalisation is a bloated myth, and they support this by showing that when one refers to an extra-national entity or level of action, the odds are that one would refer to international rather than globalised and therefore a-national entities. See Luhmann's, 1982c, concept of *world society*, a considerably more sophisticated description of the subject matter of globalisation.

in banality in both cyber and global city discourses, is only a feeble systemic attempt to erase the perennial novelty, to overshadow the space of ignorance, where absence becomes embarrassingly present. Novelties such as cyber and global dimensions, are seen, and correctly, as non-novelties by the systems. But this is a defence against the absence inside. It is systemic domestication, cunning but ultimately transparent.

The commonality between the urban and the environmental legal system lies in precisely such a self-preserving systemic decision to domesticate its environment, to deal with 'novelties' as non-novelties, to absentify absence, rather than abandon itself to acknowledge novelty, ignorance, absence. But this is not always successful. The continuum is ruptured and reveals novelties that cannot be accommodated either as novelties or as non-novelties – and this does not only happen at the full moon, when monstrous hybrids emerge, but at all times, when the system deals with its risks, its novelties, its memories. At such moments of intensification, the system proceeds into the presentification of absence, the acknowledgement of the thing that is neither new, nor already of the system, the 'other' system determined by its relation of continuum/rupture to the system. The boundary of the environment with the system is repeated in the system, not as an institutionalised and deparadoxified presence via a domesticated and appeasing external self-reference, but as the *absence* that keeps on tipping the urban form off-balance. The internalisation of the continuous rupture with the environment within the system is not a Luhmannian re-entry of the unity of difference, but an attempt by the system itself to institute a space within its boundaries that even the uncertainty following re-entry cannot populate. The difference between re-entry and continuum/rupture becomes obvious if put in terms of uncertainty/ignorance. Uncertainty is a disguised novelty, further disguised as non-novelty; ignorance is every-time-once new, always present in its absence, already *in* the system, never *of* the system. Thus, to return to the urban system, the environment of the city is its ignorance, an absolute novelty that has always been around. In its various terminologies as nature, cyborgs, teletopias, coyote, werewolves, hybrid networks, etc., the environment is an internalised absence of knowledge, linked and simultaneously alienated from the system by means of a form balancing while undercutting continuum and rupture, luring the system into (self-)knowledge, while confirming the limits of the system's ability to know.

V IN ABSENCE

After seven days of walking through forests, the traveller who goes to Bauci arrives at the city even before he sees it. The fine pillars that rise from the ground in large distances one from the other and vanish into the clouds support the city. One can climb up the stairs on the pillars. The inhabitants seldom appear on the ground: they already have everything they need up there and do not need to come down . . . There are three hypotheses for the inhabitants of Bauci: that they hate the Earth; that they respect it to the point of avoiding any contact; that they love it as it used to be before them and that they never get tired of observing it with binoculars and telescopes, leaf after leaf, rock after rock, ant after ant, contemplating fascinated by their own absence.[164]

Thanks to a felicitous wordplay in Italian,[165] the inhabitants of Bauci manage to reflect on their absence *and* on the Earth as it appears in their absence. The Baucians have been endowed with the ability to observe both the space their absence delineates – where they would normally be, had they been there – and the space around that space – the space around where they would be, had they been there. The Baucians insert themselves in the environment but only as an absence, and this is what they focus on. By removing themselves from the Earth, the Baucians define the Earth and themselves *in absentia*. What is more, their selection cannot be verified: the other possibilities are that they either hate or love the Earth, but we do not know what the case really is.

Focusing on absence is one way of conceptualising the form continuum/ rupture. Observing the unobservable can only mean observing one's inability to observe. The emergence of limits and limitations occurs at its most forceful in the autopoietic definition of the city. Cities are the walled spatiality of Cain against the unknown environment, be this God, other humans, or nature.[166] But, however high the walls, they still contain what is inside, which is never straightforwardly different to what is outside. Take the example of a *kare-sansui* Japanese garden: a representation of a cosmology that includes both itself and the larger order of nature, while at the same time situated in the midst of busy urban settings.[167] The garden is the city: continuum and rupture undercut each other and contextualise their paradox in spatialities, temporalities and corporealities that cannot promise a synthesis. 'The Japanese garden stands out as a metaphor for a primeval world *before* human intervention.'[168] The garden is the absence of the city within the city, but this absence is not vacuous: it is a definition of the environment even before it could have been differentiated from anything else. The environment surrounds nothing but its absence: a continuum ruptured by its own continuum.

Just as the Baucians insert themselves as absence and manage to define the

164 Calvino, 1993:77.
165 The final phrase of the passage reads '. . . *contemplando affascinati la propria assenza*', literally translated 'contemplating fascinated their own absence'.
166 'God the first garden made, and the first city Cain' in *The Garden* by Abraham Cowley (1618–1667).
167 Crowe, 1997:15.
168 Crowe, 1997:15, my emphasis.

space of their absence and the space around their absence, urban autopoiesis defines itself (the city) and the space around itself (the environment) in absence: by removing itself from the environment, the city manages to define both its absence and the space from which it is absent. By inserting the absence of its environment within its boundaries, the city loses the battle for definitions and abandons itself to a languid walk on its wall.[169]

169 'The trajectory taken through the city is guided by that city's voids, by what is not there. The city is constructed by its gaps' (Barber, 1995:13).

Chapter 3

Couplings

Horizons, exclusion, justice

This chapter brings together environmental law and the city. Autopoietically, the endeavour presents difficulties of various degrees, mainly attributed to the Luhmannian divide between society and the rest, as well as the concept of communication which applies only to social systems and thus reinforces the societal boundary. While this would not be so much of a problem, since there are autopoietic tools with which society and consciousness can converge (instances of which have been shown in the previous chapter), it becomes significantly more problematic when the environment returns as the main focus of the discussion. The presence of an absent environment within the system demands several shifts of emphasis and reconceptualisations that affect both the way the system understands itself and couples with other systems.

The first concept to be scrutinised and cast in the light of absence is structural coupling *– namely the historical co-evolution of two systems through mutual observation. Structural coupling is now described from the 'unfocussed' angle of the environment, which converts the already considerable levels of uncertainty produced into unmalleable spaces of ignorance within the system. This ignorance, along with the previously encountered descriptions of ignorance in the law and the city, are operationalised in a negative way through what I call* unutterance*, namely a vehicle of self-questioning with regard to the logocentricity of communication. The operationality of unutterance relies precisely on its negative qualities as non-fitting, non-accessible, non-communicable, non-essentialist, and so on. The paradoxical nature of such an operation – the* disruption *of operation as operation – is the closest a system has to a 'core', namely a space within the system that guarantees the unity of the system through its very absence from the operations of the system. This is not a tangible space, before and beyond operations, but the very paradox of operations. In its centripetal non-essentialism, the paradox is the crux of recursive self-legitimacy and the space on which the system must never tread at risk of disintegration. The paradox relies and reinforces an undecidability which is here seen to be the condition of an internalised but undomesticated exteriority. Through the form continuum/rupture, the environment of the system in its spectral appearances as justice, utopia, unutterability, returns as the absence that haunts the system.*

I HORIZONS OF COUPLING

In attempting to determine the relation between environmental law and the urban system, one observes connections of interdependence on several levels and of varying importance. On a microlevel, environmental law regulates the city by ways of planning regulations, transport, air pollution thresholds, clean water levels, noise pollution, land uses, recycling, etc., while the urban system, in its capacity as the geography with the greater concentration of environmental problems due to the density of activity, challenges the limits of environmental law and raises issues that require both localised and globalised information and decision-making procedures. Notions such as 'urban sustainability', 'urban agenda 21' and 'greening cities' come from policy-making sectors, whereas other primary tools of environmental law that come from the core of environmental legal developments, such as Environmental Impact Assessment, Integrated Pollution Control, Market-Based Mechanisms and so on, are also transformed and adapted to the exigencies of the urban environment.[1] On a macrolevel, the relation between environmental law and the city echoes the well-debated question whether law is conducive to social change or rather social change precedes legal reform. Forms of attempted social or legal coercion can be observed in the relation between the legal and the urban, especially in issues such as social justice, urban minorities, public participation, environmental degradation as a consequence/ co-requisite of social exclusion, and so on. In their ramifications, these issues offer a fertile ground for identifying the relation between the legal and the urban as being variably one of confluence, conflict, mutual influence, antagonism, constructive co-existence, control, and so on.

Autopoiesis departs from the above ways of describing connections between systems in that it does not have the conceptual vocabulary for hierarchy, power structure or even mere influence.[2] This does not mean that the existence of the above is denied. They are simply relegated to a different level, that of second-order observation. Thus, causality is replaced by *attribution*, namely the causal link performed by a second-order observer in their attempt to explain the way things work together. Attribution 'does not penetrate the units, but it can establish that they occasionally combine, that they adopt the same or complementary values for many variables, or even that they operate as a unified unit on specific occasions'.[3] On a first-order, systemic (self-)observation level, however, there can be no cause and effect. Everything is organised on a flat, unhierarchical array of systems with *their*

---

1 For the terms see Haughton and Hunter, 1994; Girardet, 1992; OECD 1990; Blowers, 1993.
2 With the possible exception of *steering*, Luhmann, 1997b, which however is never seen without its limits.
3 Luhmann, 1995a:223.

environments. There can be no direct contact, influence or regulation between systems. Power is powerless in autopoiesis because it is divided between the system and its environment. Anton Schütz puts it with poise: '*Arkhé* (mastery) is located outside and in front of the system – that is, just beyond the system's borders with its accompanying other or heteros.'[4] The system cannot govern itself because its *arkhé* is located in its environment, which is inaccessible to the system; and the environment cannot govern the system it encircles because the environment is inoperable. 'The dream of a self-mastering society, of societal autarky or sovereignty, is incompatible with the autopoietic division, the autopoietic disentanglement, of a system which is contingent but gifted (gifted with selectivity, preferences, self-reference, use of time), and an environment which is calmly indifferent but unmanoeuvrable.' And further, 'one is liable to limitless enslavement and cannot take control of its conditions. The other side fails to be elevated to a corresponding plenitude of power.'[5] This is one of the hardest paradigm shifts autopoiesis initiates, one that requires leaving behind 'the fiction that played the role of the consolatory companion throughout many centuries of European Dasein'.[6] The diffusion of power in autopoiesis is a balance between a blind potency and a luminous impotence. The environment, although predestined never to be mastered, is entrusted with the mastering of the system. The irony is that the environment has no means of awareness of this operation – it 'does' everything *in absentia*.

Ways of escaping this counter-intuitive mapping have been consistently suggested by Teubner and others.[7] The suggested approaches do not re-institute the possibility of direct regulation. On the contrary, by building on practices of observation, reflexive theorists offer a more 'accepted' version of what has been here called 'reality' – namely, an institutionalisation and management of the surprise effect without denying the basal incommunicability between systems. In this subtle way, reflexive law departs from the Luhmannian pronouncement that autopoiesis does not prescribe, only describe, and suggests ways in which the descriptions can be put into use. While to some extent following the example of departure from dogmatic autopoiesis, as it will become obvious, my quest is somewhat different from reflexive law's applicational terrain. In what follows, the connection between environmental law and the city will be seen in the light of environmental absence, namely a negation that constitutes the first but unavowed point of reference, and which, faithful to its nature, lacks both demarcation and determination.

Putting together environmental law and the city is a prima facie easy operation, even for autopoiesis. The mechanism readily at hand is *structural*

4 Schütz, 1996:275.
5 Schütz, 1996:261.
6 Schütz, 1996:275.
7 Teubner, 1988, 1989; Paterson, 2006; Paterson and Teubner, 1998.

coupling.[8] The concept has been employed both by Luhmann and Teubner, and has been heavily criticised as the demise of autopoiesis,[9] mainly because of its 'cop-out' quality from autopoietic closure. However, things are not that elementary. Structural coupling has already been introduced in biological autopoiesis as the 'history of recurrent interactions, leading to the structural congruence between two or more systems'.[10] In the introduction to *Autopoiesis and Cognition*, Beer analyses how structural coupling comes about: '[system] and environment operate as independent systems that, by triggering in each other a structural change, select in each other a structural change'.[11] The term 'interactions' must be read in close connection with that of 'triggering', since the system 'classifies and sees in accordance with its structure at every instant' such 'interactions'.[12] Structural coupling involves no contact, no intersection, no input–output. What at first may be understood as information exchange, soon turns out to be a translation of environmental perturbations into the system's own code. Nothing enters the system that does not already belong to the system. Structural coupling is simply a coupling of structural operations: the operations may emulate one another but the systems remain distinct. For Luhmann, structural coupling is an 'interlocking of independent units'[13] which contributes to the constitution of an autopoietic system since 'it presupposes . . . not yet fully determined possibilities'.[14] Thus, structural coupling 'fixes, through the structure of an emergent system, how these open possibilities are to be used as meaning'.[15] From these open possibilities, the structurally coupled system constructs its expectations from its environment: '[structural coupling] is simply the specific form in which the system presupposes specific states or changes in its environment and relies on

---

8 Luhmann has dropped interpenetration for structural coupling, thereby moving to an arguably more insular way of understanding the way systems can come 'together'. Initially only systems of the same kind – conscious or social – could couple. Indeed, this was the main difference between structural coupling and interpenetration, as discussed in the previous chapter. In the later version, structural coupling can also take place between dissimilar systems. See also King and Thornhill, 2003. I differentiate between the two, reserving interpenetration with its extended corporeal and spatial connotations for the dissimilarity of the urban system, and coupling as an indication of a certain systemic historicity, for the law and city connection. To employ the division that La Cour, 2006, proposes, in structural coupling the system employs its environment as information. In that sense, I will interpret structural coupling as the system's ability not only to respond to information (La Cour, 2006) but also, and significantly, as the system's ability to respond to the lack of information. See also Ch. 2 n. 67.

9 Munch, 1992.

10 Maturana and Varela, 1992:75.

11 Maturana and Varela, 1972:xix.

12 Maturana and Varela, 1992:74.

13 Luhmann, 1995a:223.

14 Luhmann, 1995a:221, which is especially relevant to environmental law, as I have explained earlier in Chapter 1 due to its incipient nature; see also Chapter 6.

15 Luhmann, 1995a:221.

them'.[16] If these expectations are betrayed, the system has to deal with *irritations*, which are also internally constructed.[17] Thus, a system couples with another when a series of irritations and a reciprocity of observation lead to the production of a commonality of structural history. Structural coupling presupposes and provides a *continuous* – as opposed to isolated events[18] – influx of disorder, which enables the system to learn and change whilst maintaining its structures.[19] Little by little, structural coupling creates its own historicity based on the operability of the 'relation': 'Once the corresponding selections are working, they acquire a tendency to reinforce themselves, given the irreversibility of time.'[20]

Structural coupling has echoes of the Husserlian concept of *pairing*, defined as the appresented other within the self. Appresentation in Husserl's terminology is to make 'present to consciousness a "there too", which nevertheless is not itself there'.[21] In pairing, alterity appears as primarily instituted and does not rely on inference, but on analogised transfer of the originally instituted.[22] But since the only primal instituting is the ego, pairing, as the constitution of the other, must always take place within ego. This is why the 'there too' is not itself there; this is also why, in structural coupling, a system only presupposes states of its environment (i.e., operations of other systems) rather than directly perceiving their existence. Still, amidst this internalisation, the construction of alterity seems to be mutual. As Theunissen remarks, pairing is reciprocal, in that ego is constructed by alter as much as alter is constructed by ego.[23] Analogous reciprocity is found in autopoietic coupling, in the form of a constructed mutual history. As a necessary precondition, however, the coupled systems have to be of an operational commensurability: in pairing, this is expressed 'as data appearing with mutual distinctness' and 'found phenomenologically a unity of similarity and thus always constituted precisely as a pair'.[24] Likewise, structural coupling, at least in its earlier version, can only take place between structurally similar systems.[25]

Even if one were to disregard the issue of the city's unclassified systemic categorisation, structural coupling would be a good way to begin, but not an

16 Luhmann, 1992a:1432.
17 Luhmann, 1992a:1432.
18 Luhmann, 2004:381; 1993b:441. This is another difference between interpenetration and structural coupling.
19 '. . . the organisation of the system is conserved while it undergoes structural changes' (Maturana and Varela, 1972:xix).
20 Luhmann, 1995a:223.
21 Husserl, 1973a:109.
22 Husserl, 1973a:111.
23 Theunissen, 1984.
24 Husserl, 1973a:112.
25 This is the main difference between structural coupling and interpenetration, as discussed in the previous chapter. See also King and Thornhill, 2003.

adequate way to conceptualise fully the relation between environmental law and the city. Structural coupling builds on a certain historicity of difference, where structures become shared without becoming common, and social systems can be seen in a co-evolutionary light.[26] Autopoiesis is of course, through and through, an evolutionary theory, faithful to principles of variation, selection and stabilisation of the system.[27] Co-evolution is a trait of structural coupling, and as such retains the notions of borrowed complexity, cognitive evolution, autopoietic continuation and normative closure. In this loose co-evolutionary sense, environmental law and the city can be described as being in a relation of structural coupling. There are, however, some aspects of structural coupling that need to be emphasised and further developed, in order to accommodate the peculiarity of the urban/environmental legal connection: first, the relation of continuum/rupture that characterises the coupling; and, second, the deep historicity, in the sense of a receding horizon of non-differentiation, which makes itself relevant through absence.

Observing law and the city together is not just an epistemological positioning. It is certainly that; but the observation allows and involves a glimpse of an ontology of systemic continuum that does not exhaust itself in the historical co-evolutionary mechanism of structural coupling. Even if of second-order, observation cannot capture the horizon of the coupling, but restricts itself to operations that explain the relation, attribute causality, invent origin and purpose, and, in short, create order from noise, respecting all along the intransparency of the very operation: the observer can never observe herself observing.[28] Every observation is a distinction between the observed (marked) and the unobservable (unmarked); even when one believes that one can observe the world, one always generates the *blind spot* from where observation takes place. A blind spot is the observer's 'back' to the world, as it were, which is to remain unobservable, unless through a further distinction that, in marking the previous blind spot, will generate in its turn another blind spot. It is the space of ontology in the epistemology of autopoiesis. In the process of observing the city and the law, the observer must position herself simultaneously 'against' and within the two systems, on both sides of their boundaries, straddling system and environment. This peculiarity is dictated by the nature of the environmental legal system which proceeds into the materialisation of environmental absence within, as well as by the nature of the urban system as located on the very distance between signs and observer, a moment of intensification between the conscious and

---

26 According to King and Thornhill, 2003, only between social (not conscious) systems can structural coupling be equated with co-evolution.
27 Luhmann, 2004:232*ff.*
28 One thing the observer must avoid is wanting to see himself and the world' (Luhmann, 1998a:111).

the social. Whoever the observer may be, the liminal location of the city as both/neither disallows the observer from resting on one observatory, obliging her to shift continuously from exteriority to interiority, fractally replicating the *durée* of the city, and continually generating blind spots that obscure her very observation of the coupling.

Likewise, law and the city couple in a way that obscures their very coupling. The element of historicity is already emphasised in structural coupling, but here, historicity is a way of refuting tautology between systems. Austin Sarat's dictum that 'the law is all over' is not just an instance of juridification of society,[29] but a veritable linguistic encapsulation of the paradox: law permeates everything, and this is the *end* of law. 'Law is over' is 'law is all over': when law becomes the other space, the interiorised exteriority of space, then law finds its limits in an act of self-obliteration. For, what is the point of keeping law separate from its environment, if law *is* its environment? The point of course is diffused: it is the system that keeps itself separately from its environment; or it is the observer who keeps the observed and unobserved separate; or it is Luhmann who keeps systems differentiated, for (an overwhelmingly prescriptive) fear of societal de-differentiation. Historical co-evolution hides the fact that there is no difference between system and environment (everything that will be of the system is already systemically determined in manners of structural fatalistic immanence),[30] and that 'co-' in this case maintains the illusion of difference while wrapping it in an illusion of identity. Emphasising that structural coupling is not just an exchange of undomesticated complexity, but also of undomesticated identity, is a step towards understanding the relation between law and the city as one of survival both through and against their collapse.

The above can be put more abstractly in terms of continuum and rupture, or tautology and difference between law and the city. Structural coupling visibilises the difference of identity by constructing a horizon of reciprocal construction, thereby pulling the coupled systems apart while offering a second-level, attributional identity. Structural coupling presupposes continuum in order to institute rupture which appears as continuum. The horizon of continuum, the primordial soup of pre-differentiation, returns and haunts the structures that couple, rendering them more ferocious in their

---

29 Sarat, 1990.
30 Maturana and Varela, 1992:96, talk about 'structurally determined systems', namely 'systems in which all their changes are determined by their structure, whatever it may be, and in which those structural changes are a result of their own dynamics or triggered by their interactions'. Selections are always already there, within the system's potentialities, but remain unpredictable. Systems are what von Foerster, 1984:201, calls *non-trivial machines*: 'while these machines are again deterministic systems, for all practical reasons they are unpredictable: an output once observed for a given input will most likely be not the same for the same input given later'.

closure, more solipsistic in their incommunicability. This rupture is the pre-condition of their coupling, not unlike Luhmann's cool observation that love 'is shown to be not a mere anomaly, but indeed a quite normal improbability'.[31] Structural coupling is comparable to Luhmann's description of love as reflexive, namely the love of the 'us' as a potential unity, an internalised projection of the self and other as seen by the other, but through the only vocabulary one knows: one's own.[32] If love is the love of the different disunity as internalised by myself, then all love is self-love, because the self includes the other (just as the other includes the self), provided that the other never even dreams of the platonic primordial continuum. But *this* would be anomalous.

I have looked at the incest performed between law and the city elsewhere, where my main concern was to articulate the production of a spatialised legal utopia which encapsulated moments of rest for law and justice – what I have called the *lawscape*.[33] The present project goes before and behind the lawscape, looking at the (imaginary) moment in which the two elements of the lawscape – law and the city – have been 'separated' in the first place. For this reason, the peculiarities of environmental law as the one that couples with the city are especially apposite precisely because of its articulated paradoxical relation with its environment as described in Chapter 1. Environmental law's mirroring onto its environment in the sense of an internalised continuum/rupture between systemic operations and environmental ignorance, informs the *invitation* to couple, in a way matched by the city's continuum/rupture between its systemic acrobatics and its environment. The concept of invitation is perhaps the defining difference between, on the one hand continuum/rupture, and on the other, the canonical Luhmannian articulation of identity and difference between system and environment. Continuum presupposes and brings about the invitation to couple, just as rupture enables and disables coupling; but their reciprocal invitation, their mutual entanglement enables schisms within the continuum and sutures along the rupture. There is no space clean from either, be this environment or significantly system, and resonances of both the ghost of de-differentiation (in this case, dissolution into tautology) and the threat of *über*differentiation (to the point of non-recognisability) can be found across the boundary. The invitation, therefore, is a means of appeasement and further exposure, a self-imposed utopian challenge that conditions the space of invitation by exposing it to further rupture.

---

31 Luhmann, 1998b:9.
32 In other words, *double contingency*: I can only imagine how the other sees 'us' and that is what I internalise. At the same time the other does exactly the same, perpetuating expectations of expectations *ad infinitum*, like two black boxes trying to communicate with each other.
33 Philippopoulos-Mihalopoulos, 2007, the introduction to the volume and my chapter on Brasília. For the general relation between law and space, see Blomley, 1994; Blomley *et al.*, 2001; and Taylor, 2006.

The concept of invitation emerges when one analyses the operation of observation as self-exposure in the form of the blind spot generated by the observer. Every blind spot is a step away from the theological vantage point of no-description, of reality. The blind spot guards observation from becoming tautological, one with itself and its object, self-annihilating. In its supreme immanence, the blind spot *conceals* any need for exteriority, for ethical calls, for transcendence. As Luhmann says, 'this imaginary space replaces the classical a priori of transcendental philosophy'.[34] The blind spot is always of the present and contingent, the external reference of every reference, the space that guarantees difference – which is Luhmann's riposte to transcendental unity. But as such, and precisely because of its immanence, the blind spot is the observer's Achilles heel, a necessary concession to the draughts of exteriority, an oblivious exposure to alterity in the form of the stare of other observers. The blind spot is the ground on which the invitation takes place – the blind spot *is* the invitation, has always been there, an immanent, contingent, invisible a priori, a protended appellation to the continuum in the environment, in short: an inviting void in the shadow of every observation where one can project dreams of unity.[35]

As rupture and continuum, structural coupling can be explained through the circularity of an ever-receding, ever-prior, ever-egological invitation to couple. In the invitation, a summons to continuum is protended. But in the priority of the other invitation, the rupture is reinstituted as the inviting distance of an undomesticated complexity that always precedes. Thus, the continuum is enabled precisely on account of the rupturing effect of the priority of invitation, and the rupture disabled and invited anew with every continuum. The conditions become even more undercutting when the specificities of environmental law and the city revisit the discussion. Thus, the invitation between these two is a noisier than usual affair because it includes and essentially addresses (always in systemic oblivion) not so much the other system, but emphatically the environment of the system.

This can be better understood if one constructs the basic structural coupling between environmental law and the city as one of reciprocal conditioning. This does not equate to direct contact or influence, but internalised observation that reveals the limit(ation)s of the system. Simply put, the city is regulated by environmental law, and environmental law tests itself in the intensification of the city. Thus, a cursory second-order observation would reveal that environmental law's compartmentalising, spatial ordering and temporal binding emerges in the city's working order (in its production of expectations). Conversely, the city's relative multipolarity and social differentiation fleshes out

---

34 Luhmann, 1994b:21. See also below, Chapter 5.
35 Invitation and its specific employment as an ethical call on behalf of alterity is of course a Levinasian formulation; see Lévinas, 1981; Derrida, 1999.

the 'material' side of the law, namely its relation to combinations of command-and-control and market-based mechanisms which are produced and tested on urban grounds – planning restrictions, environmental regulations, zoning, social control, consumption of resources, waste collection and disposal, borders between private, public and restricted access areas, pavements, roads, traffic lights, metro barriers, flow of people, power architecture and landscaping, are just a few examples of the way the conditioning takes shape. This is encouraged by the urban spatial intensification, in the sense of physical and social proximity.[36]

The invitation to conditioning takes place on several grounds: a generalised urban crisis, both because of intensity of spatio-temporal productions and consumption, and because of the sheer quantity of urbanised areas in the world, circularly produces and is produced by environmental regulation.[37] While the world will never become an ecumenopolis[38] or a megalopolis,[39] the tendency towards understanding legal localities in urban terms marginalises any legal signification of the non-urban, except of course if the countryside reaches the symbolic through protest – but even that will have to take place in the city. At the same time, a call for transnational environmental law to deal with urban environmental issues on a global level runs parallel to globalisation's self-legitimising mythology.[40] Whatever the ramifications of such a move, the point remains that the city is increasingly 'inviting' legal categorisation from an environmental law that finds itself less and less capable of dealing with what is expected of it.

The reciprocal protension to conditioning reveals, not a causal relationship between what is needed and what is provided, but a simultaneity of environmental internalisation as fractured continuum in both systems. Since the reciprocal invitation comes even before the need to invite, the always-already coupling of the law and city is so obvious that it becomes invisible. This is the reason why it appears impossible to be aware of the frequency of the legal presence in the city, and conceptually strenuous to think of law's materiality as formed in its urban grounding. The a priori nature of the interweaving means that there is no identifiable origin, cause or indeed discernible direction of perturbation. It becomes obvious from the above instances of invitation between the two systems that what is in play is something at the same time within and outside each system: environmental law constantly redefines its boundaries when dealing with the city, while the city constantly endures attempts by environmental law to have its precise nature pinned down. The

---

36 Bauman, 2003a; Valverde, 2006.
37 Harvey, 1996a; Potter and Lloyd-Evans, 1998; McAuslan, 1985, 2003; Hardoy et al., 2001; Gilbert and Gugler, 1992; Nivola, 1999.
38 Doxiadis and Papaioannou, 1974.
39 Gottman, 1961.
40 Mitchell, 1997; Castells, 1996; Sassen, 1994.

'failure' can be attributed to the increasingly uncertain boundaries of both the law and the city when it comes to distinguishing themselves from their environment. In that sense, the two systems keep themselves busy with each other, when in reality what they do is call for their environments to be revealed. The two systems invite, not each other but each other's environment – and this invitation does not take place through observation but its flipside, namely the production of blind spots. The city exposes its environment to environmental law and, simultaneously, environmental law exposes its environment to the city. Each expanse of ignorance becomes conditioned by the horizons of the other, in an attempt to mirror, not a secure coupling between systemic operations, but an uncoupling of ambiguity, a rupture of the unknown as totality, and a 'compartmentalisation' of this unknown into system-specific unknowabilities, which, however, become disrupted by the other system's continuum with its environment. Through the coupling of environments, the uncertainty of the environment is not just a matter of exchanged re-entries of the difference between system and environment, but a much more radical and thus uncontainable experience of exposure precisely because of the environment's nature as both rupture and continuum with its system. One, therefore, sees that the language of structural coupling proves inadequate to describe the ignorance produced by the simultaneous clashing and melting of horizons, unless redirected towards the other side of the boundary: the coupling of environments. If structural coupling has the co-evolutionary effect of cognitive development, the coupling of environments has the effect of an exponential increase of ambiguity and arbitrariness within the system in view of the amount of environmental perturbations to which each system is now exposed. To recall Calvino, '. . . his answers and objections took their place in a dialogue already formed on its own in the mind of the Great Khan. That is to say, between the two it made no difference whether they uttered questions and solutions aloud, or whether each one carried on pondering in silence. In fact, they were silent, their eyes half-closed . . .'[41] By shifting the emphasis from the system to the environment, boundaries become co-extensive with the horizon, and systems can be observed in their pragmatic (rather than differentiated) operation of having to protect themselves against the uncontainable wave of environmental ignorance. If the observation of structural coupling focuses on the way environments are being shared and imploded, horizons revealed and fractured, and absences replace domesticated presences, then one can better contextualise the trouble systems have in maintaining their autopoietic identity.

To sum up, structural coupling between law and the city is more accurately described as coupling of environments, in view of the systems' structural and operational ambiguity with regard to their environments. The environment is

---

41 Calvino, 1993:25.

the total horizon of each system, which is radically conceptualised as the continuous redefinition of continuum/rupture between system and environment.[42] When one system invites the other in coupling, they offer their respective blind spots (namely, what cannot be focused on, when the system is in focus) as ever-receding horizons of prior invitation, thereby conceding a priority to the ignorance of alter that can never be domesticated – unlike their own environment which has been converted into a domesticated external reference, or, as Baudrillard would say, to something 'dangerously similar'.[43] This is not just double re-entry, in the Luhmannian sense of systemic exchange of difference between system and environment, but precisely its converse: when environments couple, their horizons clash and collapse into each other, thereby producing levels of ignorance for the system that cannot be converted into systemic parlance, yet continually perturb the system. These configurations push systems into the internalisation of an exteriority never before encountered as such: to recall Lyotard,[44] the secret *hôte* of the system returns in the guise of counter-differentiation alliances, transcending the self-assured systemic closure and, while relying on it, rupturing it by revealing itself in the horizon of the invitation by and to an unassimilated exteriority. It is precisely this paradoxical quality of the exteriority as rupturing while continuing, that necessitates a discussion on the means with which the exteriority manifests itself on the border between ignorance and unassimilation. Indeed, this exteriority does not belong to the system's horizon proper, nor can it be converted by the system into meaning; yet, it manages not to remain ignored, and to be registered as a perturbation by the system. Luhmann does not have a term for such a concept. I will therefore call it *unutterance*.

II UNUTTERANCE

The term 'unutterance' is obviously built on a negation of utterance. For Luhmann, utterance is one of the three elements of communication (the other two being information and understanding),[45] and refers to the conversion of the distinction of information into something linguistically intelligible. Admittedly, Luhmann allows utterance to occur outside language through other forms of social communication, such as staring, dressing, signs, gestures and so on – what he calls 'indirect' communication.[46] However,

---

42 This departs from Luhmannian identity/difference in that now the focus is the environment, with its impossibility of being focused on. This is also how the formulation differs from Gadamerian, 1989, fusion of horizons.
43 Baudrillard, 1993:129.
44 Lyotard, 1993 – above, Chapter 1.
45 Luhmann, 1989b:143.
46 Luhmann, 1982a, 2000a. See also Murphy, 1984, and King and Thornhill, 2003.

all these forms presuppose language in the role of the other side of the binarism: indirect communication is meaningful only as language replacement. Otherwise, it cannot operate autopoietically; it is effectively excommunicated. In his treatise on art (a system which for Luhmann embodies a possibility of communication outside language), Luhmann makes the connection between language and communication quite clear: '[a]rtistic communication distinguishes itself both from communication that relies exclusively on language and from *indirect communications that are either analogous to language or unable to secure the autopoiesis of communication*'.[47] When King expurgates Luhmann from structuralism by mentioning that communication for Luhmann is not produced by language, he only manages to save him from the strict Saussurian structuralism and not from the encompassing structure of the social supra-system.[48] This system dictates its limits through verbalisable processes, namely events that can be expressed through language or its absence, always defined in contradistinction.

Unutterance is a shift of emphasis analogous to the emphasis given to the environment when talking about structural coupling. It is the opposite of language/non-language binarism, in the sense of going beyond the identity/difference discourse and attempting, just as the coupling of environments, to address issues of communication (not in the Luhmannian sense) that precisely transgress societal boundaries, accommodate different configurations such as the urban system, and manage to express the production and consumption of ignorance as ignorance, in the manner enabled by the coupling of environments. In other words, unutterance's etymological negation is the tool through which the negation of the system as the locus of observation materialises. Unutterance is the vehicle of the blind spot, the unity of difference that remains incommunicable.

Unutterance can initially be understood as silence, except that it goes beyond the linguistic semiologisation of silence. Silence is usually understood as the absence of sound. The usual metaphor for silence is that of a forest, where trees represent the sound, and silence the space between trees that remains largely unnoticed, but operates organically in understanding the forest.[49] Patten, in his attempt to teach students about silence, shows that the easiest way to understand silence is indirectly and in contradistinction to language. He asked his students to attend poetry sessions where poetry was read aloud and where they should attempt 'to hear a diversity of silences in different voices' and 'to forget about the poem's meaning'.[50] The introduction

47 Luhmann, 2000a:52, added emphasis; in that sense, unutterance is not far from art. The implications for law are only baffling.
48 King, 1993.
49 Clifton, 1993.
50 Patten, 1997.

of silent, as opposed to oral, reading that addressed an audience, has disqualified the importance of what I would be tempted to call 'sonorous' silence, or else, the silence between the sounds – the spaces between the trees. We know that silent reading was practically unknown to scholars of classical and medieval worlds, and, even during the Renaissance, the word 'reading' connoted oral reading.[51] It is noteworthy that we have partly been deprived of the appreciation of silence through the introduction of silent reading. Indeed, silent reading has covered up the multiplicity of silence, which is the hiding place of polyphony, and has replaced it with one sonorous unity. While there is nothing inherently problematic with it, it has a problematic consequence: silence has been populated by sounds, and more specifically by words. One is satisfied to define silence antithetically, as the absence of language or of specific sounds. To borrow de Sousa Santos' words, silence is no longer conceptualised as an amorphous infinite 'but [as] a reality delimited by language, as much as language is delimited by silence'. Silence is 'the self-denial of specific words at specific moments of the discourse, so that the communication process may be fulfilled. In that way, what is silenced becomes a positive expression of meaning'.[52] And although this is an achievement in itself, silence remains what language is not: one can only see silence when one sees the trees – without trees there is no silence.[53]

The above is in line with the autopoietic dealings with silence. Silence is an opportunity for society to explore its as yet unprescribed folds. In his article *Speaking and Silence*, Luhmann describes silence not as an operation outside society 'but only a counter-image which society projects into its environment'.[54] According to Luhmann, silence – just as anything else – can only be seen through society. Silence remains communication. Society does not recognise what cannot be verbalised because it does not know how to deal with it. In silence, society sees only the environmental threat, the ignorance of what cannot be communicated. Silence, according to Luhmann, 'is the mirror in which society comes to see that what is not said is not said'.[55] Not that what is not said *cannot* be said, neither that it is said but not *understood*, nor that there is *no point in being said*. Society cannot accommodate its ignorance and prefers to convert into communication whatever little it can understand – hence the communicative effect of gestures, looks, etc. There is little point (other than linguistic) in arguing that Luhmannian society cannot operate in

---

51 Pugh, 1978. One reason given for the decline of oral reading is the increase in readers and book availability; another, more convincing but sadly no longer valid, is the hum under the dome of the British Library.
52 Sousa Santos, 1995:150.
53 Silence can be meaningful both literally and metaphorically: see Jarowski, 1997 and Sczuchewycz, 1997.
54 Luhmann, 1994a:33.
55 Luhmann, 1994a:33.

any other way except on the basis of language/non-language.[56] For even non-linguistic utterances are defined in contradistinction to the other side of the binarism and are filled contextually with their mirror image that lies on the marked space: what would this silence mean in language?[57]

Unutterance goes beyond silence, in that it does not purport to be communication in the societal way, it cannot be converted into language or meaningful gestures or even absence of that, and it is not a projection of society. Unutterance remains a meaningless perturbation irritating the system through its *malgré soi* invitation. It lies beyond the communicative possibilities of meaning, and cannot be domesticated by a system. Unutterance is not simply a retreat from communication, something that 'no longer wants to be understood as communication (but is forever understood, is understandable only in this way)',[58] but a double negation: negation of language and negation of non-language.[59] By underlining and erasing the form language/non-language, unutterance silences the logocentric meaning of silence. As negation, unutterance is situated within the system: it is a reversion from the inside, a meaningful negation (in the sense that it results in meaninglessness) from within, that interrupts the domestication of the system and brings embarrassment at the non-connection.[60] Just as coupling of environments does not connect in the way structural coupling does, unutterance does not connect meaningfully, nor does it contribute to the autopoiesis of the system in the way operations do. It is certainly generated by the system, produced

56 'Connecting language with the social dimension produces the concept of communication' (Anchor, 2000:108).
57 This is particularly relevant to law. Peter Goodrich, 1987:56, writes: 'Language is the first institution held in common; it is the universal law prior to Babel; it is the invisible writing of law in the heart, prior to writing; it is the only inscription of law which can escape the idolatry of other signs.' In its textuality, law leaves little margin for the symbolic, let alone for the unsymbolisable. Goodrich, 1990, refers to the case of the Haida Indians who chose to support their claims to their native land through full ceremonial dresses and masks, not lawyers, but armoured with tellurian mythologies, traditional poems and heroic songs that, for them, demonstrated beyond any doubt their ancestral claim to the land. Unsurprisingly, the court decided that their claims were not legally relevant: the court refused to accept exteriority as incommunicability and classified it as meaningful but nonsensical in terms of law. Goodrich suggests that the court refused to compare mythologies, because that would raise questions of the 'self', of what it is that the court represents. It is through an unutterable exteriority that the system faces itself – see also section IV, below.
58 Luhmann, 1994a:27.
59 Luhmann, 1995a, following Gotthard, talks about the *rejection value*, namely the value that does not fit the code except as a destabilising perturbation – and in that sense, unutterance is a rejection value. The difference lies in the invitation of unutterance, that populates the absence with an operability.
60 'Within a meaningfully self-referential world organisation, one has at one's disposal the possibility of negation, but this possibility can only be used meaningfully. Negations, too, have meaning, and only thus they connect up with anything' (Luhmann, 1995a:62). This negation, however, is undomesticated, meaningless, irritating and disruptive.

internally as a result of the blind invitation through its environment, but it is there as an active disruption of autopoiesis, a hole of incommunicability, a suspicion that there is something else there, something that does not partake of the altercations between language/non-language. The system's autopoiesis is framed by this stronghold of internalised ignorance, a *memento vanitas* of the system within the system, a perpetual source of uncertainty, anguish, internal perturbations; in short, a valuable stimulation that manages to communicate nothing to the system but its very absence of meaning.[61]

The connection between unutterance and environment is positively ambiguous. In its decontextualised form, unutterance stands for the presence of absence. When this is transposed specifically to environmental law and the city, unutterance brings forth without defining the continuum/rupture between system and environment. Absence is populated by uncertainty, ignorance, the outer side of the boundary, the systemic fear in the invisible face of alterity. As absence, it can never be understood by the system as part of its present autopoietic operations, nor does it represent a space of cognition, since it has already been, as it were, 'rejected' yet imposed. Absence remains present, immanent, and intentional, parallelising without connecting inside and outside. This means that unutterance cannot represent a side of things, nor can it be employed as the one who speaks for the silenced.[62] It may be employed like this by an observer, but its operation is less pronounced than that. Unutterance is the disruption of the system in a way that the system accepts, precisely because it 'appears' as a result of the latter's environmental coupling: an exteriority that remains so, while revealing itself (always as incommunicability) within the system. Unutterance is populated by the fear of the system for its continuum with the environment, and creates within the system a space at the same time alluring and prohibited, a *nothingness* that pushes and pulls by clogging the operations of the system while being of the system. In talking of nothingness, one is reminded of the Heideggerian 'the nothing' (*das Nichts*) which appears always as a 'correlative' of being,[63] the precondition of knowledge, which, however, remains unacknowledged.[64] For Heidegger, *das Nichts* is always of this world, always interfolded with the world and never taking

---

61 There is a mysticism in unutterance not dissimilar to the mysticism of intentionality: in Pentecostal religious communication, God's answer to the prayer comes like a reverberation within the praying individual, an inner dialogue where recipient and sender are interfolded into one being (Sczuchewycz, 1997). The word of God is communicated to the believers as silence. Luhmann, 2001:18, admits that religious silence is no longer communication. Celestial intentionality is little more than the fear of phenomenological bracketing. But unutterance is more than that: it is stopping before the limit, a pause that silences and petrifies.

62 Cf. Cornell, 1992b; Spivak, 1989.

63 Heidegger, 1992.

64 Heidegger, 1996.

one outside the world (an immanent, rather than Husserlian, transcendence). And in so doing, 'the nothing' reveals being: only through 'the nothing' can being be revealed, and then, always attached to it, paradoxically united in a difference that cloaks it and makes it what it is, and returns it to where it belongs, always 'held out into the nothing'.[65] The form continuum/rupture appears here as the impossibility of defining one (being, system) without interrupting the other (nothing, environment), and for that, without the other. But 'the nothing' is not just negative – in fact it is emphatically not the *nihil absolutum*;[66] but always the one side of the form which, through its absence, envelops presence and makes it probable.

In another famous Heideggerian text, the form appears in its absolute inoperability as the 'Black Forest farmhouse'. This typically Germanic space is the abode of ideality, in which 'earth, heaven, divinities and mortals enter *in simple oneness*'.[67] Oneness expresses the fearful continuum between system and environment, as seen in both environmental law and the city – in other words, the systemic prerequisite of differentiation from an exteriority that must remain so. Nature, along with humans, is located outside Luhmannian society; but unlike humans, it merits no other mention, except as the instigator of 'too much resonance' in society.[68] Of course, nature *is* a construction, and its description is mediated by this very fact. For this reason, nature is accepted, either as a Luhmannian domesticated, constructed systemic 'resonance'; or, more radically, as an absence presentified in the systemic fear of continuum with the environment. But the point remains that there is no plausible way in which to understand the continuum with plants, ecosystems, the earth, the universe.[69] This is the continuum between the first ('natural') and second ('artificial') nature observed in Marx by Murray Bookchin,[70] Adorno's identity between first and second nature,[71] as well as Bateson's unity of interiority and exteriority.[72] It is there where unutterance enters, as the system's own open aporetic wound, as the guest whose hostages all systems are,[73] and who, without coercion, force, aggression, but with an unspeakable urgency and immediacy mediated by its very blind invitation, nests within the system and nags it into a co-existence with its blindingly visible absence.

65 Heidegger, 1996:103.
66 Heidegger, 1992.
67 Heidegger, 1971:160; Heidegger has been widely referred to in deep ecology, not always with a positive outcome. See Foltz, 1995, and Zimmerman, 1993.
68 Luhmann, 1989b.
69 This is Eckersley's point, 1992, when she dismisses Habermas's communicative rationality, 1987, because of its inability to include anything but human beings.
70 Bookchin, 1990b.
71 Adorno, 1973.
72 Bateson, 1988.
73 Derrida, 1999.

This 'positive' use of negativity is, of course, not novel.[74] Recently, Bruno Latour has employed the impressionistic negativity of 'controversy' in order to initiate the debate between the two 'collectives' that do not communicate with each other: humans and non-humans.[75] His use of *speech impedimenta* is comparable to unutterance: 'The first speech impediment is manifested by the multiplication of controversies: the end of nature is also the end of a certain type of scientific certainty about nature.'[76] Uncertainty should be considered an inevitable ingredient of environmental crises, and as such should be accepted rather than attempted to be clarified with more knowledge. Acceptance of uncertainty replaces something that cannot be debated with something that can – and this includes what Latour calls *speech prostheses* invented by humans in order for non-humans to participate in the discussions, when the former are perplexed as to the position and utterances of the latter. Such prostheses 'give worlds the capacity to write or to speak, as a general way of making mute entities literate'.[77] Latour returns to a logocentricity that relies on a belief in consensual debate, with the unwanted consequence of appropriation of alterity within a rational, political system. Admittedly, Latour tries to avoid this by placing an emphasis on the cross-fertilisation between humans and non-humans, but it is difficult to see how he does it outside the limitations of *logos*. This is where unutterance departs from the positivisation of speech impedimenta, and proceeds to an immured uncertainty that blocks *logos* and makes it question its operations.

Unutterance brings in an absent environment and covers it with a visibility beyond systemic intelligibility. Unutterance remains non-consumable: a geometric aide, an imaginary line that helps solve the problem in hand, a tool that vanishes as soon as it is used. It also remains non-producible, echoing what John Cage has said in referring to silence: 'try as we may to make a silence, we cannot.'[78] Unutterance cannot (and does not need to) be 'produced' and 'inserted' in the system – after all, nothing could be 'produced' and 'inserted' in the system; it is already in, despite the system, in systemic ignorance, reconstructing an immanent asymmetry and creating pockets ready to receive ignorance. It is an echo of the systemic environment in its partial totality (all the totality that the system can and cannot understand – but this is not the total totality, just the probable systemic totality). Epistemologically speaking, to accept unutterance as the means of non-communication within

---

74 From Hegel to Heidegger and Adorno, negativity has been used in various ways; most relevantly in Adorno's *Negative Dialectics*, 1973, whose inversion involves a positivisation of the excluded 'object' as a means of fighting against traditional Marxism's use of nature. See also the beautiful *Pas* by Derrida, 1976b
75 Latour, 2004a.
76 Latour, 2004a:63.
77 Latour, 2004a:66.
78 Cage, 1961:8.

the system means placing a little more confidence in the system than Luhmann does. Optimistically or plainly naively, absence of communication (in any formulaic combination with non-communication, interaction, perception, and so on) is a re-entry of the difference system/environment, but this time with the environment paradoxically present. This is no longer just a shift in emphasis but a radical reinterpretation of autopoiesis as an unfolded form whose paradox is employed because of its inoperability. The acceptance that systems operate counter-operatively brings forth a rather normal operation of the system, which has been overshadowed by the emphasis on communication and its dogmatic separation from perception – in that sense, observation discovers its 'real' nature. By reducing one's observational scope, by focusing on the boundary and forgetting big ideas such as society, communication and reality, one is able to let the other side emerge in its quietude and claim its quite normal disruptive role.

III PARADOX

With unutterance, one hits upon the best-diffused secret of autopoiesis: the paradox. Luhmann defines paradox as that which 'wants to use simultaneously what is incompatible and thereby deprives itself of the ability to connect'.[79] In its inability to connect, unutterance institutes a space of delirium, an internalised rupture which becomes operational *because* of its inoperationality: unutterance subverts not only logocentricity in the narrow sense, but *logos*, and with it the rational and logological structures that serve the systems' cognitive openness.[80] Unutterance introduces a disorientating space within the systemic structure which cannot be approached via systemic operations. A cancerous formation of sorts, produced by the very system in its invitation to ignorance, which wraps the absence of the systemic environment in a cloak of irritating presence: unutterance expresses (by not expressing anything) the void of ignorance that goes so deep in the system's structures that it manages to remain absent from the system's normal operations, yet underlines them as a permanent threat of implosion. The co-existence of absence and presence at the same moment as one event that undermines itself by not generating meaning or connections, is a *form* before its rupture, an unmarked

79 Luhmann, 1994a:26.
80 In that sense, unutterance is not dissimilar to art: 'Like indirect communication but in different ways, art escapes the strict application of the yes/no code ... In avoiding and circumventing language, art nonetheless establishes a structural coupling between the systems of consciousness and communication' (Luhmann, 2000a:20). See also Philippopoulos-Mihalopoulos, 2004c. This is why, according to Withers, 1997:351, 'silent' works of art appeal to the spirit rather than the mind. See also Hafif, 1997, who points out that silence in art is often described as monochrome.

totality without a space from which to be marked. Forms are paradoxes in that their connection is one of paralysing transgression, entirely closed and self-referential: absent *because* present, and present *because* absent.[81]

Etymologically, paradox is 'the other belief' (*para-doxa*). Paradox is the first and last instance of dialectics. It initiates the other speech, the other speaking, their expressing a belief contrary to the belief of their interlocutor. This contrary belief, however, is equally valid, with the result that the discussion returns to itself without ever concluding anywhere: neither anywhere outside the *doxa* and the *paradoxa* (say, a third *doxa*); nor anywhere within the initial dialectics. Instead, it carries on whirling between the two initial *doxae*, without ever reaching a conclusion. But this irritates a dialectics habitually guided by reason, and (paradoxically for autopoiesis) begets a 'transcendental necessity'[82] for some sort of solution – not necessarily synthesis, but an end to the paralysis, inability to communicate, inability to distinguish, no sense, nonsense.[83] Thus, the need of deparadoxification.[84] Luhmann identifies the following modes of deparadoxification: 'unfolding, making invisible, civilizing, making asymmetrical',[85] all four emanations of one primordial operation: that of *distinction* (decision, asymmetrisation, marking, prioritisation).[86] Paradoxes must be deparadoxified, because otherwise they remain inoperable. The system cannot operate with forms: only by marking itself against the environment, or by marking the observed against the unobservable, can the system deal with unity. Unity expresses an omnipresent impossibility, both alluring and threatening in its fragile perfection. And perfection is already there, achieved, reached – thus, ready to be disrupted. Paradoxes need to be kicked out of their smug utopia and into an operable (but still utopian) 'reality'; they need to be severed in order to serve the system. But even when severed, paradoxes return in myriads of ways and haunt the system from within, check on its readiness to deal with uncertainty, or its consistency when dealing with itself and its boundaries. The paradox holds a mirror in front of the system and places the latter *en abyme*,[87] an

---

81 Luhmann, 1986c:395*ff.*
82 Luhmann, 2000a:132, remarkably referring to paradox as a 'transcendental necessity'.
83 See Goodrich, 1999; Teubner, 1997 and 2001a.
84 There is nothing untoward about this: every man kills the thing he loves, etc. For an expansion on deparadoxification based on a confluence between Derrida and Luhmann, see Philippopoulos-Mihalopoulos, 2003. See also the entire collection on paradox, Perez and Teubner, 2006, and especially the contributions by Teubner, Clam, Perez, and Kastner.
85 Luhmann, 2004:64. In the original, 1993b:23, the phrase reads: '*Entfaltung, Invisibilisierung, Zivilisierung, Assymmetrisierung*'. Also Luhmann, 1989b:144–5. One could comment on all four of them and amuse oneself especially by questioning the ramifications of the third, since, interestingly, 'the paradoxicalization of civilization has not led to the civilizing of paradoxicality' (Luhmann, 1990a:134).
86 See e.g. Luhmann, 1998a:108*ff.*
87 Derrida, 1984.

infinite game of mirrors where discrepancies fractally repeat themselves and disrupt the system from the inside.

Unutterance stumbles on exactly such a paradox. I will call this paradox *unutterable*, because of its impossibility of being uttered without the system's risking disintegration. An unutterable paradox is the application of the paradox onto itself: is the paradox paradoxical? Is the law lawful? Is the city environment? Am I me? The supreme unutterable paradox in autopoiesis seems to be that of the world: 'the world cannot be communicated and, when the world is included in communication, it appears as the paradox of unity of difference, a paradox that requires a solution [*Auflösung*] if things are to continue at all. In this case the world remains incommunicable.'[88] While unutterable paradoxes radically interrupt communication and bring operations to a standstill, *utterable* paradoxes abound and are consistently coped with. Utterable paradoxes are fairly innocuous forms, managed by the system in a routine way. Thus, the very identity of the system is defined in the paradoxical opposition between hetero- and self-reference, quickly deparadoxified in the absorption of every identity description as self-description.[89] Normative closure and cognitive openness can co-exist attributionally because of the 'open because closed' formulation.[90] Likewise, observation exists because of its blind spot. Specifically in the legal system, the form of norm/interpretation is a classic utterable paradox, also referred to by Luhmann as core and 'periphery'. Courts are to be found in the former, in contradistinction to other kinds of legal communication, such as the legislature, that correspond to the 'periphery' of the system.[91] As expected, Luhmann was quick to dispel any impression of hierarchy between core and periphery, since hierarchy would be a misinterpretation of autopoietic circularity.[92] The courts are simply the elements of the legal system that deparadoxify law's utterable paradoxes, not least because their operation itself is based on a constant expectation of deparadoxification (that of the legal obligation to decide);[93] but also because courts are the guards of systemic memory, the facilitators of its evocation. This means that the form norm/decision is deparadoxified with the help of conditional *programmes* (namely, computation between an *if* and a *then*, or else an (external) fact and an (internal) legal rule)[94] through

---

88 Luhmann, 1994a:27.
89 Luhmann, 1992a.
90 Luhmann, 1988b:20.
91 Luhmann, 2004. One can see that this is a turn from the usual continental centrality of the norm, to a more common law-friendly judicial centrality: compare with Luhmann, 1985a, where such division does not appear. See also below, Chapter 6.
92 Luhmann, 2004:277.
93 Luhmann, 2004:292.
94 Luhmann, 2004; also Luhmann, 1989a, and 1989b; for a comprehensive application, see King, 2002.

differentiated prioritisation between what can also be referred to as the internal and external sides of the form. The same internal/external differentiation can be witnessed in another Luhmannian binarism, that between *redundancy* and *variation*.[95] Variation is the systemic accommodation of surprise, whereas redundancy is akin to the memory of the system. While there is little doubt that the system accommodates surprises according to its memory, variation triggers not only evocation but also new combinations – in other words, intelligence. Redundancy, on the other hand, is the process of banalisation of external perturbations that takes place in strict accordance with the system's memory. The system is expected to balance both functions without compromising either its ability of cognitive openness to innovations, or its structural unity. In balancing, the system takes into consideration two kinds of consequences: the intra-systemic consequences which refer to future legal decisions, and the external consequences, or the effects a decision has on the legal environment.[96] It is not as if redundancy and variation have respectively internal and external consequences, or that the legal system can consciously select which mode of reaction it will employ. Rather, the connection is one of contingently balancing one binarism against and through the other without any prioritisation – otherwise known as a paradox.

Likewise, the urban system consists of utterable paradoxes. Its very identity as an oscillation between social and conscious, and communication and interaction, is generated by a contingent and 'external' prioritisation as a result of intensification. At the same time, urban space is produced *because* it is consumed, and consumed *because* produced, managing to define its topology at every instance without these coterminous processes losing their distinctiveness. On a different level, its ontology is relativised and at the same time materialised by the city's location in the sphere of 'reality', thereby unfolding the paradox through the absence instituted by utopia. The urban system consists entirely of utterable paradoxes, without any supra-structure limiting them or directing them, nor with any reliable, predictable continuity of identity (which is made observable only on account of an adequately developed, in-built ability to accommodate surprise).

Every system abounds in utterable paradoxes, while keeping itself away from the 'big' unutterable. Utterabilities are the marked side of the form utterable/unutterable, where the latter can never be marked, although is used as an external reference. Every utterable paradox is a prohibitive reference to the unutterable, thereby taking the system away from the unutterable and into the quotidian practicalities of communication. In every paradox, the unutterable is found *under erasure*, never to be found, on a level at the same time simultaneous and non-simultaneous to the utterable. 'Simultaneous' because

---

95 Luhmann, 1995b.
96 Luhmann, 1995b:294.

it is constitutive of the form; 'non-simultaneous' because it can never be evoked while the utterable is speaking. The communicative utterable keeps the unutterable at bay. At the same time, every utterance of a paradox confirms the relevance of the unutterable by visibilising its invisibility and maintaining its inoperability.

Paradoxes are the *élan vital* of autopoiesis, the continuous questioning of an untraceable origin, the Derridean *trace* of the cardinal distinction. Paradox keeps autopoietic systems capable of maintaining their autopoiesis without relinquishing their immanence. *Paradox is autopoiesis*, in all its lithe claustrophobia.[97] However deep into the system and further from its boundaries one looks, the paradox as the fear of fraudulent function can never be totally erased. On the other hand, the more precarious a system's identity, the more flexible its dealings with paradoxes. The less 'seriously' a system takes its self-description (that is, the more imbalance there is between its external and self-reference), the less threatened it will feel from its unutterabilities. This is not an abdication of the responsibility that comes with functional differentiation – on the contrary, it is the embracing of the infinitely more demanding responsibility of changing, of adjusting to expectations in ways that may compromise assumed levels of external reference.[98] In other words, the unutterability of the paradox is the call from the entrails of the system to look 'out', to search for its boundaries and reinforce them in a way that would include its very unutterable aporia. Paradoxes do not always have to be unfolded, as Luhmann prescribes. Increasingly, one observes systems being momentarily paralysed by the utterance of the unutterable; systems that despite themselves find themselves exposed to vociferous articulations of their unutterable paradox; systems that are found coupled with other systems, and whose environments become intermeshed and their horizons unsettling; systems that drift in the immured spaces of unutterance and happen upon the arbitrariness of their differentiation. In such cases, the system will eventually unfold its paradox and will try to avert its stare from whatever threatens it, either by passing it over to other systems, or by diffusing it in gregarious utterabilities – there is no doubt that autopoiesis needs to carry on. However, there is a moment of 'exchange', where the system fumbles for its boundaries, an instance when its space of ignorance reverberates in the system's self-description with a vocabulary that problematises its code, where the unutterable paradox slides upon itself and threatens the system with an ultimate tautology between itself and its environment. Such

---

97 This point goes somewhat further than Jean Clam's, 2001 assertion that 'without paradox, there is no autopoiesis', in the sense of the Derridean use of the cupola, resonant of 'deconstruction is justice' in Derrida, 1992a, and the Lévinas-inspired 'intentionality is hospitality' in Derrida, 1999.

98 Ladeur, 1999; see below, Chapter 6.

instances, although isolated, are increasingly becoming catenations of rupture, allowing descriptions perilously to approach self-descriptions, causing problems to the systems and taxing their habitual mechanisms of dealing with uncertainty.

IV IMMANENT ABSENCE

The unutterable paradox is distinct for each system. Luhmann talks about the application of the systemic code onto itself – thus, the question whether law is lawful is the fundamental paradox of the legal system.[99] However, unutterable paradoxes share a common characteristic regardless of their systemic particularity, namely, their inclusion of exteriority. Underneath the question whether law is lawful or unlawful lurks the potentiality of an appeal to the outside of the system, to an arbiter who is better equipped to judge than the system. In posing to itself its unutterable paradox, the system questions itself and its authority with regard to its identity, and this is why the system needs to make its paradox visible. The lawfulness of the law has been intimately connected with violence and the latter's legitimating force, until the moment that law expunged and replaced this external reference of violence with self-reference, thereby covering up the paradox: 'whenever violence is involved, the paradox of legal coding shows up – but in a form which is immediately unfolded within the legal system through setting conditions which make the paradox invisible'.[100] These conditions are legally defined and allowed violations of the law – hence, the shift of the reference from external to internal legitimation. But even in the most successful internalisation, the original fear of continuum between system and its environment is poorly hidden. Equating law with violence, although far from original,[101] is the original exteriority of law, the fear of disintegration because of an inclusivity that threatens the system's very self-description, its idea of societal function, its purposelessness.

The exteriority of the city is another instance of the fear of continuum as the materialisation of the unutterable paradox. The city's exteriority is everything that the city is, yet refutes – its monsters, its signs, its waste, its risk, its all-encompassing colonising presence, its environment and its absence. The city fears its environment (with all its resources that make up the city) as much as it fears its inner folds of all-inclusivity. The unutterable paradox of the city is for the city to be tautologous to the very thing that it destroys: its environment, its construction of nature and countryside, the semiology of the

---

99 Luhmann, 1989a and 2004.
100 Luhmann, 2004:265; see also Luhmann, 1985a, and Rogowski, 1994.
101 E.g., Weber and Benjamin meeting in Derrida, 1992a.

non-urban (whatever this may be), the narratives of a different city, the uto-pia. Through its oscillating systemic nature, the city manages to invisibilise the paradox of its identity as a self-imploding totality, thereby hiding its exteriority (exactly like law) as internalised constructions of self-suspension: the city gates are open to everyone, as long as they make themselves scarce in the face of a system that takes pride in its inclusivity. The city is the only place where riots can take place and be quashed, the 'subaltern' can find an audi-ence and be further marginalised, nature can be 'effectively' represented and colonised, movement can be pushed to its conceptual extremes and intoler-ably restricted. The unutterable paradox is the beast of self-dissolution in the face of exteriority that appears, after all, in a continuum with the system. But is exteriority really exterior, is it simply something internalised and domesticated, or is it to be found in the whispers of unutterance?

To clarify this, I return to law and to an impressionistically different way of looking at its exteriority, namely, no longer as violence but as justice. Luhmann locates justice at a radical distance from systemic coding: 'The traditional question on the justice of law loses all practical meaning.'[102] The passage from natural to positive law falsified any theoretical impression of continuity between law and justice,[103] since such accounts could not account for the presence of, as well as need for, complexity. Accordingly, justice is a *supra-programme* of the legal system, namely a criterion that applies to the code lawful/unlawful and determines its selection, leaving the gates open for a filtered complexity to be understood by the system. Justice 'guides' legal decisions from outside of the process, as 'a programme of (all) programmes on the level of the programmes of the system',[104] or as a *circular formula of contingency.*[105] The criteria for justice are to be found within law: '[s]ince only positive law is "valid", namely able to use the symbol of legal validity, one must not look for criteria outside ("*rechtsexternen*") but within law ("*rech-tsinternen*")'.[106] There are two things worth noting about these criteria: first, the legal system does not employ them in order to be just, but to be consist-ent: 'these criteria become relevant through the question as to how, in the face of the increasing complexity of law, it is possible to continue taking consist-ent decisions (that is, distinguishing between cases that are alike, and ones that are not)'.[107] Second, the criteria may well be inside the system, but justice itself as a value remains outside the legal system, in its environment, at a safe

102 Luhmann, 2004:212.
103 Luhmann, 2004.
104 Luhmann, 2004:213.
105 This circular formula appears on the level of programmes. Circular formulae are self-referential modes of crossing the boundary between determinacy and indeterminacy; see Luhmann, 1990d:879*ff.*
106 Luhmann, 2004:225.
107 Luhmann, 2004:225, footnote omitted.

distance from its code.[108] Of course, the socially understandable expectation of continuity between law and justice demands an internalisation of *some* idea of justice within the legal system. When the issue of fairness of a legal decision is posed, the legal system can only go back to its set of criteria for the application of its code lawful/unlawful (rather than just/unjust) and adjust accordingly. However, it will never know whether its decision has been just or unjust. The system's internalisation of justice – in the form of a supra-programme or a contingency formula – is a domesticated construction of justice's exteriority, which, however internally convincing, cannot distance itself from its fundamental fallacy: justice remains beyond the legal system, in an exteriority whose continuum with the system has been ruptured for fear of endangering the universality of the systemic code. This can only mean that the legal system *may* be just. What is surprising about the last statement is how little surprising it is.

Justice is an exteriority that the law cannot understand. Luhmann has no issue with this – in fact, he bases the whole idea of legal validity on it ('legitimacy is based on a legal fiction')[109] as the only way in which the system can deal with complexity. But the problem remains, however circular and self-referential law's legitimation may be. The noise of complexity is becoming louder and louder, and exteriority demands to be heard. As Jiri Pribáň puts it, 'the social space of instability and constant confrontation is the space in which legitimation takes place . . . [N]oise is not only contingent, but also subversive.'[110] The subversion of exteriority is an anathema to the law, which armours itself behind its codified application of justice and presents itself unable to grapple with its unutterable paradox: its continuity with an exteriority that is not accepted in the system except in the noiseless shape of contingency formula. But how can this carry on, especially when the law is questioned by its own bodies of appearance: take environmental law, which consistently deals with a noisy, demanding, intrusive uncertainty, whose ramifications are often extra-legal arbitrariness, disillusion with the traditional forms of legal protection and increasing experimentation with hybrid forms of transdisciplinary construction of expectations. Environmental law is only an example of a system's – any system's – unrest with its own self-description, and its fear about its possible continuum with other, more critical, more incomprehensible, more relentless descriptions. The unutterable paradox is repeatedly murmured from numerous sides and in numerous ways, and the suspicion enters the system's night sweats.[111] It is no longer an issue of

---

108 The terms 'just/unjust' are not to be found in the vocabulary of the legal system, for otherwise justice would have to be added as a third term to the existing code; Luhmann, 2004.
109 Luhmann, 2004:123.
110 Pribáň, 2001.
111 See Luhmann, 1995e; also Goodrich, 1999.

theoretical description, but of practical prescription. It is the gap between law as it describes itself, and law as it is being described. It is the rupture between law's operations and law's theatricality, the symbolic forum of law where 'the social drama of affect is acted out in resilient theatrical forms', as Peter Goodrich reminds us.[112] It is the gradual corrosion of law's self-description as the institutionaliser of normative expectations, and its embarrassment before an overabundance of cognitive expectations.[113] It is the theory observing itself changing, softening, pluralising.[114] It is also the autopoietic description's unwillingness to take into consideration its unutterabilities as vociferously whispered in legal theory.[115] To put it in terms that the legal system could not understand (but everyone else would), it is a question of why law is not just.[116]

The question of locating the exteriority returns. Outside, it cannot be heard. Inside, it becomes domesticated. Jacques Derrida has offered his own aporias to this, which I would like to visit briefly in order to allow for their positional exteriority to flow into the text. Derrida begins by distinguishing: 'law (*droit*) is not justice'.[117] Justice is an elusive, incalculable chimera that can only be negotiated outside the law. Justice, like democracy, is 'always untenable at least for the reason that it calls for the infinite respect of the singularity *and* infinite alterity of the other as much as for the respect for the countable, calculable, subjectal equality between anonymous singularities'.[118] Still, the impossibility of justice is the condition of its possibility, and for this one needs to strive. Striving for justice takes place through the 'closest to what we associate with justice', namely law. Legal 'calculations' are the way to reach justice, while simultaneously underscoring the distance between law and justice. As Howells puts it, 'it is the mismatch between law and justice . . . which is the very condition of justice'.[119] This mismatch is complicated, as much as facilitated, by the 'ghost of the undecidable': '[t]he undecidable . . . is not merely the oscillation between two significations or two contradictory and very determinate rules, each equally imperative.'[120] Instead, the undecidable

---

112 Goodrich, 1999:212.
113 See earlier, Chapter 1.
114 Cf. Luhmann himself, especially post-90s; Gunther Teubner *in toto* as a valiant ambassador of autopoietic pluralism; Ladeur, 1999.
115 '[T]he theory of autopoietic systems powerfully relativizes the practical significance of legal dogmatics and even more so the practical significance of legal theory' (Luhmann, 1986c:409).
116 Teubner, 2001b:35, looks at the gap between the legal and the external through what, reluctantly, he calls 'a chance for another re-entry', or, following Luhmann, a partial self-transparency of the legal re-entry. The obvious connection with Derrida's position of justice is explored by Fitzpatrick, 2001. See also Philippopoulos-Mihalopoulos, 2004a and 2005b.
117 Derrida, 1992a:14.
118 Derrida, 1994:65. One is reminded of Luhmann, 1989b, who considers democracy a fortuitous improbability.
119 Howells, 1998:152.
120 Derrida, 1992a:24.

is, once again, the elusiveness of justice when it comes to a decision, either because without the undecidable it may simply be legal, or because, after the moment of undecidability has passed, the decision will no longer be *'presently just, fully just'*.[121] Indeed, '[t]he undecidable remains caught, lodged, at least as a ghost – but an essential ghost – in every decision, in every event of decision'.[122] Derrida sees the 'ordeal of the undecidable' as the condition of the im/possibility of justice. Although wedged in between the events of a decision, the undecidable is the gate to the horizon of probabilities, the values-that-have-not-been-chosen-but-could-have-and-still-can. But such realisation of environmental alterity can only be thought of as taking place exclusively *through* and *in spite of* the self (the system): 'through' because it is through the internalisation of undecidability (or the horizon of probabilities) that the self can appreciate alterity; and 'in spite of' because it is only through interruption of the self that alterity could be materialised.

Derrida's positioning of justice bears obvious similarities to Luhmann's. Most fundamentally, justice as contingency means that justice is probable, or at worst improbable, but certainly not impossible. In other words, justice is introduced in the horizon of probabilities of the legal system as one option amongst others, and this is what privileges it over impossibility. Thus, Derridean im/possibility are the two sides of the Luhmannian selection. In this respect, the undecidable of the event of the decision is on the same level as Luhmann's operation of in/determinacy in the event of the foundational decision of whether a case is like or unlike other cases.[123] The undecidable/ indeterminate has no origin within law: Luhmann alludes to an a priori quality of justice when he says that 'the norm of justice must be accepted without knowing beforehand which decisions will follow from it and which interests it will serve'.[124] Indeed, the two approaches share the same basic schema:[125]

---

121 Derrida, 1992a:24.
122 Derrida, 1992a:24.
123 Luhmann, 2004.
124 Luhmann, 2004:221.
125 This does not mean that the two theorists agree. Differences abound, and can be found not only in the two texts in question, but most significantly in their total oeuvre (see indicatively Fuchs, 2001; Luhmann, 1995e; Teubner, 2001a; Cornell, 1992a; Philippopoulos-Mihalopoulos, 2004a). In the present context, the main difference is the following: while for Derrida justice seems to be the ultimate, albeit chimeric, purpose of the legal system, for Luhmann, justice is not an indication for the direction of the system in the sense of need for more justice: it cannot be 'understood as formula for development or as an indication of the desired direction of a system's development' (Luhmann, 2004:222). Luhmann deals with the lack of direction in the form of internalised self-reference. Thus, there is no purpose in the system except for the purpose of the system, also to be found in the system. The difference with my position is that I argue for a 'beyondness' of purpose, which, however, as I analyse elsewhere, 2004a and 2003, remains circular because it applies mutually exclusively to both law and justice in the form of reciprocal suspension or interruption.

justice is beyond law but operates from within law, *through* and *in spite of* law. The combination of within (*through* law) and without (*in spite of* law) is indicative of the way exteriority works in the present text. Justice is seen as something different to law, a formula for contingency whose understanding and checking goes beyond the coded abilities of the law, a ghost of the undecidable that can never coincide with the law, but employs law as the basis of calculation through which justice can be targeted. At the same time, justice is within the law, enmeshed in its operations, shadowing its decisions in ways that the law can neither understand nor avoid. It is obvious that justice is to be located in the space of unutterance as opened/closed within the system, an immanence that remains necessarily absent, for otherwise it could never irritate the operations of the system. Unutterance is the way the system 'deals with' something that it does not understand, but without trying to understand it. It is not a compromise, nor a resignation; it is plainly the limitation before the limit. Unutterance is accommodated by the system as the invited guest, the opposite of communication/interaction, the black hole that allows the draughts of an irreducibly complex exteriority to gust into the system and disrupt it, wedge between its operations and clog it with doubt and unutterability.

But is this the ethical call for the system? Is unutterance comparable in effect to Drucilla Cornell's judge, or William Rasch's Lyotardian archipelago?[126] Luhmann's position on morality is, as Rasch, correctly I think, characterises it, prescriptive: since it threatens functional differentiation by imposing a parasitic code of good/bad on differentiated systemic codes, morality should be replaced by *ethics* (as morality's reflection theory), namely the societally incorporated encouragement of distinction, of making choices, reinforcing thus the functionally differentiated identity of the sub-systems.[127] Accordingly, 'good' is the continuation of distinction, something that cannot be guaranteed by morality's propensity to be used 'in the name of . . .' with known catastrophic results; but it can be guaranteed by ethics' reflection on morality, through the promulgation of the fragility between contingent choices. This is especially the case in environmental issues, where morality enters the communication of anxiety regarding ecological crises and blurs distinctions. In Luhmann's admission, environmental ethics has not (yet) developed as a reflexive theory of morality, and still allows ecological communication to be guided more or less by moral considerations. This, for Luhmann, is a problem – a problem, what is more, that cannot be solved painlessly with the usual recourse to pure social communication. It is worth quoting at some length from Luhmann's final page of *Ecological Communication*: 'In the case of morality and ethics the concern, naturally, is with a

126 Cornell, 1992b; Rasch, 2000:199*ff*; see also below, Chapter 6.
127 Luhmann, 1991; also 1989b and 2004.

social regulation, but precisely because of this we will have to ask whether the conditions and forms of this regulation do not have to change if they are extended to an unrelated domain, to non-social sources of problems . . . An entirely new dimension of complexity comes into play through the difference of system and environment, and it is improbable that this complexity could be transferred to [habitual social] conditions.'[128] In other words, exceptionally for ecological issues, societal communication may need to be adapted to this new dimension of complexity, since habitual societal mechanisms are proving inadequate. What is more, such an adaptation involves something even more radical: 'This could give us pause to wonder whether it is not the recognition of paradox that is the way for ethics to do justice to the new problem situation – for, even in the case of theories, a more complex problem situation changes the conditions of adequate internal complexity. It could very well be that the digestive as well as the ruminating apparatus of ethics will have to be equipped with more stomachs – above all with one for paradoxes.'[129] The paradox is allowed in the systemic operations as a way of increasing systemic complexity, on par with environmental complexity. This time, the paradox is not sought to be unfolded, but 'stomached' whole, in its formulaic inoperationality. This, I think, is the event of unutterance: its ability to 'stomach' the paradox.

Arguably the unutterable paradox of unutterance is the co-existence of inclusion/exclusion in a single stillness. Unutterance cannot claim to talk for anyone. It does not aim for an inclusion without exclusion, of the critical kind so challengingly criticised by Luhmann.[130] Nor does it attempt to solve the paradox of democratic exclusion.[131] Unutterance superimposes its own paradox on existing political, legal and ecological paradoxes and brings the system closer to the memory of the unutterable paradox, without however exposing the system to it. The paradox of unutterance cannot be resolved, and its straddling between system and environment renders it omni*absent* and inoperative. The operability of unutterance lies not in its ethical stance, nor its revelatory potency. On the contrary, and rather unspectacularly, unutterance announces nothing and causes nothing. But, just as the Lyotardian *differend* cannot be resolved, only felt as such,[132] in the same way unutterance

---

128 Luhmann, 1989b:142.
129 Luhmann, 1989b:142. Luhmann refers specifically to the paradox of the moral code and its relation to ethics, but he deliberately employs a higher level of abstraction to include a comment on the need for deparadoxification in view of hypercomplex problem situations such as that of the environmental crisis.
130 Luhmann, 1995f:140*ff*.
131 Politics may *think* that it is the system that can deal better with societal exclusion, but in reality it can do little more than any other system, in view of the differentiated power allocation of functional differentiation. Luhmann, 1995f, as well as King and Thornhill, 2003.
132 Lyotard, 1994:234.

is 'felt' inside as the odd thing in, the *sublime* that marks the limit, the absence that haunts: as Rasch puts it, 'with the sublime, we do not have an observation of the excluded, but a "feeling" of, and for, the mechanism of exclusion'.[133] Inside, yet pointing outside, not by pointing but simply by being inside, unutterance is the expression of the impossibility between the practical and the theoretical, or the quotidian and the ethical: it is the bomb that blows up the bridge, the Deleuzian war machine that you invite home for tea.

Luhmann's discovery of his own 'Tristes Tropiques' was to happen in the Brazilian favelas, where exclusion sneaked into inclusion, like a pocket of entropy within order.[134] Ever since he started visiting Brazil and observing its society, Luhmann began assembling his own measured apocalypse to his theory. In some respects, he simply carried on with a project of self-doubting, the seeds of which can be found in the above extracts of *Ecological Communication*. In brief, functional differentiation is being threatened, not by its ghost of de-differentiation so much as by a superimposition of the code inclusion/exclusion on society as a whole, rather than on isolated systems.[135] This means that, while exclusion has so far been habitually dealt with by each system separately, with this new aperture to the world of ecumenical exclusion from society as evinced in the favelas, exclusion is exponentially transferred to all functional systems rather than being limited to one. This is described by Luhmann as an anomaly, a sort of 'transient condition of development',[136] an extreme case where even law is seen as 'a sheer instrument of power'.[137] It is not important that Luhmann isolates exclusion. What is important is that it is being referred to as an anomaly of functional differentiation *included* in the operations of differentiation. Luhmann talks about a 'negative integration' – a contradiction in terms and operations which, additionally, is 'nearly perfect'.[138] This, of course, is none other than the perfection of the paradox, of a form that describes the world and itself without allowing for any space of observation: an all-swallowing act of self-indulgence which, in its universality, manages to appear invisible, to irritate the system from within and to keep society awake by asking the all-too facetious question: 'Are you asleep?'

Upon waking, we see no one. The environment of the system appears within the system through a vanishing act. The environment remains an

---

133 Rasch, 2000:208.
134 Balke, 2002, first made the connection between the two, as well as a suggestion for reintroduction of space in autopoiesis; see also Diken, 2004, who, only marginally following Luhmann and mainly through cinematographic and critical references, offers a construction of the favela as a space that both conditions and escapes the social.
135 Luhmann, 1995d and 1995f; see also Luhmann, 2004:488*ff* and Braeckman, 2006.
136 Luhmann, 2004:490.
137 Luhmann, 2004:110.
138 Luhmann, 2004:489.

immanent absence, never there in the sense of codifiable information, always there in the sense of a haunting echo of the unutterable paradox. The system has invited the environment's blinding apparition without its knowing, through an invitation that is prior to the need of inviting. This is certainly a way in which to cheat the system – except that it is the system's way (hence, the only way). It is also a way (perhaps also the only way) in which the system can internalise its ignorance and at the same time freeze it as ignorance, thereby simultaneously bringing it closer and pushing it away. In this pushing/pulling act, the system familiarises itself with its limits and limitations, and wakes up to the fact that, although there is no way of knowing what it cannot know, at least it can know that it cannot know. The inversion, from the system's internal side and reference, to the system's invited space of exteriority (in other words to the environment of the system in its apocalyptic absence within the system), rides on a temporal slice of intentionality linking (without prioritising) system and environment, and looks at autopoiesis in a slightly imbalanced, slightly delirious, slightly cheating, slightly optimistic but probably 'realistic' light.

V IN *FLÂNEURIE*

What happens when a second-order observer observes the unutterable paradox of a system? Not much. There is a distance between the philosophical attitude of second-order observation and action.[139] A second-order observer constructs their own observational cosmos from their own observational point, trailing attributions and instituting horizons. Any transition to first-order levels of observation result in a deficient transfer of knowledge and only impressionistically approach action. A second-order observer is too busy not dealing with their own paradox to be able to change anything around (or in) them. A second-order observer, in either systemic or conscious guise, is a *flâneur*: the quiet and mostly invisible resistance to the operations of the observed society from within.[140] This metropolitan ideal as the space simultaneously outside and within modernity, canvassed the city as 'a theatre of new, unforeseen constellations',[141] 'a series of stages'.[142] At the same time,

---

139 'The actor's mode of attribution (first-order observation) is distinguished from that of the observer's (second-order observation). While the actor finds the bases for action primarily in the situation itself, the observer sees the actor-in-the-situation, looks for differences in the interpretation of the situation by different actors and makes attributions primarily in terms of the personal characteristics of the actor' (Luhmann, 1989a:25); see also, Luhmann, 1998b, Chapter 3.
140 Benjamin, 1983.
141 Benjamin, 1985:170.
142 Harvey, 1989:5.

however, the *flâneur* is on that very stage, incarnating the fear of and desire for continuum/rupture. Thus Baudelaire: 'his passion and profession are to become one flesh with the crowd. For the perfect *flâneur*, for the passionate spectator, it is an immense joy to set up house in the heart of the multitude . . . in the midst of the fugitive and the infinite . . . to see the world, to be at the centre of the world, and yet to remain hidden from the world.'[143]

The simultaneous distance and immersion in the urban spectacle constitutes the paradox of the flâneurie. In constantly consuming and producing spatial meaning, baptising the city, pacifying the Freudian *Unheimlichkeit* of the unfamiliar space and converting it into a chartered place of observability, the *flâneur* exposes herself, allows her blind spot to lay agape, loses herself in what she observes.[144] The *flâneur* finds herself on both this and that side, exposed and invisible, easy and restrained in her movement. In her body, the *flâneur* concentrates the systemic fear of the incommunicable. In a controlled space such as the city, the *flâneur* is not always welcomed.[145] The city attempts to bring her forth, populate her exposure with intelligible communication, masticate her absence with traffic lights and private gardens, legalise her movement with tickets and passes. The *flâneur* balances precariously between presence and absence, city and extramuros, law and justice, reality and utopia, linking without linking, describing without prescribing, enlightening without naming. In her perambulation, the *flâneur* embodies the surprise of 'reality' that remains supremely indifferent to the possibility of communication, retaining the extraneousness of the continuum and abandoning the holy battle of enlightening oneself and others. No moment of transcendence there – at least not a different one to the one that has preceded the abandonment. Just the maintenance of a steady flow of surprises. Thus autopoiesis: conscious of its inability to change the world, autopoiesis hides its inability worse than other theories. Autopoiesis proclaims only its description. But descriptions vary, and so do their receptions. Only thus can a theory change anything – by avoiding itself and its paradoxes while indulging them.

A *flâneur*-like balance between immersion and distance from the unutterable paradox fills unutterance. It appears as an invisible space of paradox that beckons without allowing entrance, nesting itself in the folds of the urban legal body and making it shift, stretch, recoil. It marks the visibility of the form continuum/rupture and the limits of ignorance with regard to whatever that human/social/natural is. It striates the Robinsonian island of Speranza with savages and castaways and brings them into a coupling of horizons that

---

143 In Baudelaire, n.d., as cited in Mazlish, 1994:50.
144 Lechte, 1995:103. Frisby, 1994; Tester, 1994:7; The *flâneur* seems to be always a 'he', which makes the female pronoun even more relevant here due to her 'invisibility': Wolff, 1990.
145 Edensor, 2000; Toon, 2000; Massey and Jess, 1995; also Sibley, 1995, on the stigmatisation of the spatially excluded.

visibilises the paradox of their continuum and rupture. And if it is true that the *flâneur* is 'an inventor of new languages',[146] in her liminal identity between selected and non-selected she can go beyond the limits of language and, if not invent new languages, at least question existing ones.

Through such an exercise of 'self-control' in the face of the savage, the Robinsonian system reinforces its unity and confirms its ability to acknowledge its limitations. But the savage is already well into Speranza, and remarkably spatial co-existence alters neither the incommunicability between them, nor the respectful simulation of each other.[147] In that sense, autopoiesis does not attempt to replace incommunicability with dialogue. It simply suggests a way for incommunicability to remain thus and still be in the system.

146 Chambers, 1994:23, as quoted in Haddour, 2000.
147 I have in mind Michel Tournier's, 1972, version of the story.

Chapter 4

Risk

Future, science and the precautionary principle

The most prominent manifestation of intrasystemic ignorance is the concept of risk as situated in an impossible to know future. Ignorance of the future is a 'uniquely common' way in which every system familiarises itself with its limitations. While future as ignorance affects all systems, the future of each system is unique to each system. This is exemplified in the way every system situates its own future within its present and attempts to deal with it on the basis of its operations. This normal inclusion of the environment of the system within the system, in the form of a divinable part of the systemic horizon, is what I will be exploring in this chapter. Operationally, the future here appears as a generalisation of the concept of unutterance, since they both are attempts by the system to accommodate its environment within its boundaries. The fact that this operation is necessarily accompanied by uncertainty is a confirmation of the inevitable concession to incommunicability as the way to approach environmental ignorance.

It is more than a word game to say that the ecological risks that the future harbours have reached the point of risk. Through the inherent threat they represent, ecological risks constantly bring social and conscious systems before bifurcations ridden with what Luhmann calls ecological angst,[1] which further affects them in extreme ways that range from technological regression to apathy. Such a situation simply confirms that risk is not merely a choice amongst options, but also a choice amongst risks, both in the sense that with every risk selection new risks open up, and also in that any risk selection as such is risky. In the past, systems used to externalise risk and attribute it to their environment, which would usually assume an extra-human guise. In view of the multi-leveled presence of risk, however, which is a relatively new phenomenon, externalisation is no longer a solution; instead, systems now internalise risk and its risks in their attempt to comprehend and prevent it from materialising. This is what environmental law does by employing the precautionary principle, a binding legal principle which operates as a form of postponement of decision in the face of scientific and environmental uncertainty. This is certainly not the only way the legal system deals with risks (another way being liability, as exhaustively presented in Teubner et al.,[2] or traditional methods such as reversal of

1 Luhmann, 1989b.
2 Teubner *et al.*, 1994.

the burden of proof[3]) but it is an appropriate example of the way the environment is included in the system. Thus, the mechanics of the internalisation of absence as witnessed in the previous chapters are going to be advanced in two ways: first, in the way the system internalises its future risks; and second, in the way the precautionary principle concretely deals with the environment of the legal system, by combining systemic closure and openness without endangering the structure of the system.

The above discussion, however, requires some preliminary remarks on the nature of time and more specifically systemic time, which will inform the description of risk as a projection of temporalities. After this, the discussion can carry on with the precautionary principle, but not conclude with it, since, as I show, ecological risk is a concept that appears in several systems at once, therefore rendering a discussion on politics and science necessary for the understanding of risk.

I RISK AND TIME

In Book XII of the Odyssey, when Odysseus prepares himself and his crew for the sailing of his boat in the vicinity of the island of Sirens, whose sweet song can be so alluring that men fall in the sea and swim to them in the full knowledge that they are going to be eaten alive by them, he chooses to command his crew to plug their ears with wax – so as to hear neither the Sirens, nor Odysseus – and to tie him up on the central mast of the boat, but leave his ears unplugged: in this way, even if he begs them or orders them to release him, they will not be able to obey. The epic encounter can be read in several ways: the most obvious one is to become part of the story, identify with the actors as first-order observers, imagine the seductiveness of the Sirens and get upset with Odysseus who decides for the crew, whereas he safely enjoys the beauty of the song. This reading echoes Adorno and Horkheimer's analysis, where typically the crew – the labourers – must concentrate ahead, blindly rely on the 'seigneur' and his oppressive role, and disregard their volition, which would almost certainly be for the taking of the risk of being eaten alive![4] Such a reading also echoes Luhmann's understanding of risk, which at least partly is based on the differentiated impact of risk between decision makers and those affected by decisions.[5] While this dimension of risk will be taken up later, at present what is of interest is another reading, one that centres not on the heroes, but the ship. The ship here, in her superficial closure, deals with risk in an enviable way: she exposes herself to risk, yet manages to emerge unscathed from it. In her constituent parts, the ship balances on the boundary between closure and openness, plugged and

3 Ladeur, 1994.
4 Adorno and Horkheimer, 1997:32.
5 Luhmann, 1993a.

unplugged ears, risk taking and risk aversion, all in one smooth sailing. How she manages this, and indeed whether the technique can be transferred to a reality devoid of mythical encounters, is what I would like to discuss here. Other readings of the myth will appear in the course of the discussion, which will shed light on aspects of risk of interest here. But first an attempt to define risk, a task which will prove even more elusive than the Sirens.

A generic definition of risk would aim at the impossible. Any attempt to define risk will inevitably focus on some measurement[6] of some aspect of either the probability of the occurrence, or the consequences of the occurrence in terms of seriousness and scope.[7] The numerical or percentage outcomes attempt to include both the probability behind 'if' and the scope of 'what', with all the intermediate combinations of 'and what if'. In other words, a definition of risk involves nothing less than the construction of a hypothetical sentence where both 'if A' and 'then B' remain unknown: a semantic equation where both variables are requested. This can be analysed conceptually in two stages: the first I would call the *perception* of risk, which refers to the definition of the specific risk,[8] and the second the *comprehension* of risk, which includes the further definitional steps that target the pragmatic positioning of a risk and will normally be in the direction of preventing risk or remedying the materialisation of risk, after having confronted it with cost, political stati, other risks, and other relevant considerations – a 'comprehensive' comprehension. The inseparability of these two stages is what lends risk its paradoxical character. Any definition of risk simultaneously tries to negate the subject matter of definition: the fear[9] of risk renders obsolete the need for a definition of risk *per se*, directing instead to risk comprehension (and thus prevention by aversion or limitation or any other way possible): we define risk by defining the initial steps of any method of prevention, which are the calculation of probability and of consequences. Thus, an adequate definition of risk would have to reach beyond a blunt statement such as 'risk is the probability of something negative happening' or even 'the probability of something negative happening multiplied by the severity of something' and establish what this probability is (degree of probability) and what this negative effect is (how negative an effect should be to qualify as negative, how and when it will materialise, further seemingly connected or seemingly unconnected repercussions, etc.), in short, well into the realm of knowledge

---

6 For a classification of different types of risk definitions based on measurements, see Femers and Jungermann, 1992, quoted in Tellegen and Wolsink, 1998; see also Harremoës *et al.*, 2002.
7 Tellegen and Wolsink, 1998:147.
8 The term does not only refer to the faculty of conscious systems, but is risk-specific and applies to social systems as well.
9 Maguire, 1996, for an account on fear and risk; Lopez, 1987, for a psychological approach; Elin, 1997, for urban-inspired fear.

of the specific risk and of its subsequent prevention. But risk is exactly what we do not know: there is little point in describing as risk something that we know is going to happen and that is going to have this and that effect. Risk presupposes and perpetuates ignorance, so by trying to define risk we negate risk. Hence we confront risk by circumventing it, we employ ostrichism and superficial euphemism to bypass it, we (think we) define risk by (thinking that we are) defying it. Risk is nothing more but the limit of our ignorance, waiting to be ruptured through intelligent planning, luck or another level of ignorance. There is nothing extraordinary in this: the definition of risk is problematic, because it is impossible to define non-risk. A situation of non-risk can only be a past situation, where the outcome is known, or a modernist utopia, where risk has been abolished in favour of a perfect order free from anything haphazard, accidental or ambivalent.[10] In defining risk, we also define risk aversion and we marry the two in the measurement of probabilities. The practical consequence of this is that we accept risk only on the condition that it will not really materialise. When we decide to take a risk we target the confined space of negation within the risk ('I will escape it'): underneath our reluctant 'yes' there lies an ever-sonorous 'no'.

This is not solely a theoretical problem: perception (definition) and comprehension (decision on action) resist separation in practice too. Thus a characteristically Cassandrian Ulrich Beck: 'society is becoming a laboratory, because everything can be checked only after it has been constructed.'[11] We lead 'destabilised' lives (if one can accept that life can ever be stabilised)[12] where a risk cannot be defined with any certainty until lived through – but then again, it will have ceased being a risk. The main example of catenation of risks, especially prominent in urban living, is technology: the more risks technology tries to prevent, the more risks it brings along. The theoretical consequence is of course that the usual binarism between risk/security no longer stands. Indeed, stating that 'there is no risk-free behaviour',[13] Luhmann proceeds to define risk not as opposed to security, but as opposed to danger. The differentiating factor is decision: risk can be attributed to a decision; danger cannot. 'Only in the case of risk does decision making play a role. One is exposed to dangers.'[14] In that sense, danger is produced in the environment of the system, whereas risk is directly attributed to the system.[15]

Luhmann's division between risk and danger is explored in one of his least autopoietic texts despite its being well into his autopoietic period.[16] *Risk*

10 Bauman, 1998.
11 Beck, 1995:104.
12 If, that is, deviation from 'normality' has ever been possible: Ladeur, 1994.
13 Luhmann, 1993a:28.
14 Luhmann, 1993a:23, brackets omitted.
15 Luhmann, 1993a:21–2.
16 Luhmann, 1993a.

preserves and develops the concept of second-order observation, but distantiates itself from the usual autopoietic considerations of closure. This becomes obvious in the risk/danger binarism, and the way that danger can be attributed to the environment, which epistemologically clashes with the environment's usual role as domesticated external reference. Nor does Luhmann institute what has been referred to here as an absent environment, namely an environment whose inaccessibility cannot be reduced to external reference. For Luhmann, risk/danger is a matter of decision making, and as such, I find that it leads away from the boundaries of the theory and into an oddly conceding attempt at empirical observation, which, however, does not fit well with the rest of autopoiesis. A way of dealing with this would be to rely on the overview abilities of second-order observation, and admit that only on that level can danger be seen as what it is, that is, external to the system, and thus imperceptible by the system. This hypothesis, however, must be rejected in view of the fact that Luhmann refers specifically to parties affected by decisions, themselves not being the decision makers, that perceive risks as dangers.[17] Even if this problem is circumvented by recourse to empirical observation, another problem remains: Luhmann himself admits that decision does not operate causally, but as an *attribution*, that is, as a second-order description of first-order causality.[18] In the sphere of attribution, the division between risk and danger becomes blurred: 'in the accumulations of the effects of decision making, in long-term consequences of decisions no longer identifiable, in over-complex and no longer traceable causal relations, there are conditions that can actuate considerable losses or damage without being attributable to decisions – although it is clear that without decisions having been made such detrimental effects would never have occurred'.[19] In other words, when causality is broken, risks become dangers. But what about attribution? What stops a second-order observer from attributing harm to a certain decision of a certain decision maker? One thinks of the obvious example of ecological catastrophes, where strict liability operates as attribution regardless of proof. The binarism becomes even more schematic if one relies on attribution, since attributions are selections of the observer, which can easily replace broken causalities.

Hence, while enlightening in its employment of decision as a differentiating factor, the binary division is not without its problems. For this reason, it seems to me more malleable and conceptually faithful to autopoiesis to accept that all risks are risks, namely endosystemic. This has been insinuated by Luhmann in a later text, where he notes that 'people are affected by natural catastrophes, but they could have moved away from the endangered area or taken out insurance. *To be exposed to danger is a risk.*'[20] Once it is

17 Luhmann, 1993a:109.
18 Luhmann, 1993a:25.
19 Luhmann, 1993a:26, footnote omitted.
20 Luhmann, 1998a:71, my emphasis.

accepted that everything that can be considered risk is risk, the obligatory differentiating binarism can be integrated within the concept of risk. Thus, risk lies within, but as a space of ignorance which, precisely because of its continuum with the present, challenges the system into cognitive explorations – into prognosis. The endosystemic location of risk renders the space of ignorance absent, and for this inaccessible. This can only mean that the system's risk comprehension can only be limited to the limits of risk, to its boundaries within and never to the risk itself – for otherwise, there would be no risk. The space of risk remains inaccessible, but precisely on account of its being within, it is perceived by the system and tickles it with its fluctuating boundaries.

Thus, there is always risk, and there can never be total security. This is conceptually comparable to Beck's announcement of risk society, whereby post-industrial society is no longer dominated by capital, but by the distribution of risk.[21] However, I would like to preserve Luhmann's scepticism on science, protest movements and other aspects of postmodernity Beck seems to favour,[22] and employ decision as a differentiating factor, although situated further along the line of risk definition. So, if everything that can be considered risk is risk, then risk should be understood as the measure of losses versus profits.[23] This measure, which entails a decision, has to be related to time and space in order for risk to be perceived as of the system.

The three temporal instances of past, present and future will have to be employed in relation to the operation of decision making. Since decisions can only be made in the present, prima facie risk is:

1. the projection of *the* present into the future
2. the projection of the present into a non-future
3. the projection of the non-present into the future
4. the projection of *a* present into the future.

Projection is not a decision. It precedes decision as a necessary stage of measurement which then forms the basis of the decision.[24] By present I intend the spatio-temporal emplacement of the system – be this the city, law or an individual – which, however, always reverts to the past 'stock of knowledge'

---

21 Beck, 1992.
22 See Harrison, 1995, for a comparison.
23 Of course, measurement as a factor of risk definition has been discounted by Luhmann, 1993a:7, as convention. However, shortly afterwards (p. 23) he talks about 'the possibility of loss occurring' when determining risk attribution, which certainly implies the necessity of measurement. In any case, convention is nothing less than the precondition of linguistic communication, especially in matters of time and space – which is what risk is about!
24 The term here refers equally to legal, political and scientific projections as well as those of the general public.

in order to acquire and organise the measurements of the future risk. Thus, present includes the past in its contextualisation, so much so that the above can be formulated as 'the present projection of the past into the future' and so on. The reason I have not chosen this, however, is because of the way the past is qualified via memory, and the relation of the past to the present, which are discussed below in Chapter 6. By future I intend a spatially contextualised time, but of unknown parameters: the present as it may be. So, the system projects what it perceives as the present into the future. Thus in 1, the system literally thrusts the present into a futurological whirl; in 2, risk is abolished immanently, since there is no future – the theoretical case of a total meltdown of the world; in 3, the connection between the system and the risk has been lost at present: again, risk is abolished, because the system does not know the existence of risk at present, so risk does not exist;[25] and in 4, the system takes into consideration alternatives to 1, and opts for one of them.

Following this reduction, risk exists only in cases 1 and 4. The difference between the two lies in the perception of 'other' present conditions. Inappropriately, this phrasing constructs a pseudo-hierarchy: it is as if the present and a present (which includes all presents except 'the' present) belong to some sort of perception-pyramid where one perceives 'the' present before the 'other' present. This image cannot be sustained: the difference is in the use of perception towards a decision. There is no hierarchy amongst presents, there is only a level perceptual field where one opts either not to decide or to decide, to remain inert or to choose. So 1 is what 4 is, but without taking into consideration the alternatives and simply letting the present develop into future as it is. And although it would seem that it is only the existence of alternatives (of 'other', of 'a' rather than 'the') that actuates decisions, in fact decision making always follows both 1 and 4, for even the decision not to decide is a decision.

At this point, a second reading of the song of the Sirens merits evoking, one that exemplifies the relation between past and future. The Sirens in Odyssey know the past, the total past, the universe and the individual. Their song is about everything that has ever happened in the macrocosmos and the microcosmos: the Sirens' song applies the vastness of knowledge to the confines of individual desire, it shows to the sailors that the immensity of human misery can be alluring if diluted in the oceans of the total knowledge. The Sirens destroy memory as past knowledge of risk and resemiologise it as beauty of the present, convert it into liquid desire, the absolute continuum with present knowledge. The risk projection of the passing sailors is therefore mutilated,

25 This is a case where the Luhmannian danger could be applied by an enlightened second-order observer. However, from the point of view of the system, there can be no risk and certainly no danger – unless the observer manages to communicate the existence of danger to the system – in which case, danger will become risk.

for if a projection that is based on past knowledge, and past knowledge as sung by the Sirens, shows that there is no risk, no rupture, only desire, the projection becomes a masquerade of the past and a triumph of the present as the sole victor. In the absence of past there can be no risk, even if the world is full of dangers. Nor can the sailors rely on anyone else to tell them that listening to the song is risky, because once listeners, they cannot acknowledge risk. And even the retrospective projection of someone who has already succumbed and died cannot be imparted to the newcomers: the incommunicability between the living and the dead is an aspect of the temporal incommunicability between systems.

Time is intrinsically connected with risk. Risk projection assumes the possibility of distinguishing between past and future in the present. This seems to be at odds with the fact that systems can only operate at present. This means that operations can only be present. Luhmann's basic tenet with regard to time is that 'everything that happens happens simultaneously'.[26] Along the Einsteinian lines of simultaneity, Luhmann contends that the environment of a system exists simultaneously with the system and that all systems are synchronised by their autopoiesis.[27] Since autopoietic systems are guided solely by their own operations, time on the operative level has no importance. Systems are guided reflexively, therefore, by their own past: 'they can gain no access to their future. Hence, they move backwards into the future.'[28] But this can only happen in the present, which, being right on a (if not 'the') border between past and future, is 'the invisibility of time, the unobservability of observation', hence 'the representation of simultaneity in time'.[29] Indeed, the invisibility of the present has replaced the invisibility of eternity as 'the vantage point from which the totality of time could be simultaneously observed'.[30]

Some qualifications on the above are needed: simultaneity of operations

26 Luhmann, 1993a:34. It is interesting to note that Gumbrecht, 2001, finds Luhmann's discussion on time conservative and almost underdeveloped especially after its replacement with contingency as *Eigenwert*.
27 Luhmann, 1995a, 1993a.
28 Luhmann, 1993a:35.
29 Luhmann, 1993a:42. Cornell, 1992a, criticises Luhmann for this 'presentocracy' with a counter-argument inspired by Derrida's *différance* on the 'not yet of the never has been', which responds to Luhmann's 'the future cannot begin'. However, Cornell's criticism of Luhmann's present-centred time concept, which is allegedly based on the modern repeatability of past and future, is not accurate: Luhmann himself, 1993, states that everything can only happen once, thereby rejecting modernity's interchangeability of past and future.
30 Luhmann, 1993a:40. This is reminiscent of Husserl's *now* as the point of actualisation of retention and protention, or else the primary recollection and the expectation of what is to come. Luhmann repeats Husserl when stating that the present is invisible: for Husserl, the now is incapable of independent existence but it is always qualified by the past and the future. See Husserl, 1991.

does not necessarily mean simultaneity of temporalities between systems. The system produces a certain temporality through its spatial positioning, its topology. Luhmann accepts that 'temporality excludes an immediate and point-for-point correlation between events in the system and events in the environment. Everything cannot happen at once. Preserving the system requires time.'[31] And preservation of the system is simply selection of one side over the other. In Luhmann's words, 'to pass from the one (indicated) side to the other, we need to perform an operation – and to do so we need time. We must cross the boundary separating the two sides and constituting the form. To this extent the respective other side exists both simultaneously and non-simultaneously. It is simultaneous as a constitutive element of the form. It is nonsimultaneous to the extent that in the operative utilization of the form (we refer to it as "observation") it cannot be used simultaneously.'[32] Simultaneity refers to a predistinction universe devoid of any ability to observe (to operate) either itself or others. This universe needs Spencer Brown's schism between the observer and the observed, the autopoietic difference: it needs a present. From the boundary of the present the universe constructs the past and the future, which it cannot observe except as present, or else, as the positioning of the observer right on the distinction: it is precisely this positioning that guarantees the invisibility of time and of present. This, however, can only mean that the distinction of past and future exists within the system, because the system does not operate with forms – that is the before distinction co-existence of the sides of a distinction[33] – but with the 'operative utilization of the form', that is after distinctions. By this I mean that the chronology of the system as perceived by the system is not to be discounted in view of cosmic simultaneity.[34] Forms are not what systems operate with. Forms exclude systems in their simultaneity. Systems operate on the basis of their own spatialised temporality – however constructed or environmentally inspired: 'if a system always had to react to environmental events the minute they happened, it would have little chance of selecting its mode of reacting.'[35] When they observe and couple with other systems, temporalities are replaced by measurements and conventions, offspring of what Maturana and Varela call 'natural drift' of the system, painstakingly produced when temporally differentiated systems find themselves coupled.

In terms of risk, this translates into intersystemic incommunicability. My past present asks me why I have been so cautious or risky, as the case may be. A future present will probably ask the same. But these presents may belong to

31 Luhmann, 1982a:292.
32 Luhmann, 1993a:36.
33 Luhmann, 1993a and also 2000a.
34 I have in mind Husserl's 'Immanent Time' as the internal time which operates as the essential
    precondition for the ultimate synthesis of consciousness. See Husserl, 1991 and 1973a.
35 Luhmann, 1995a:186.

other systems as instances of their own chronologies: thus, my future risk may not be a risk at all for the system, whose past present has already experienced the outcome of the decision on risk. Causalities may still be inaccessible, but at least the system will know what has happened. But this cannot be communicated to me, because it is still into the future. The other cannot know how I will interpret and accommodate risk, even if the other had access to my guiding selections. Incommunicability among systems is a result of the particularisation of forms. When it happens, factual communicability is all the more surprising because of the added difficulty of time co-ordination. Whether communicability, in the form of simultaneity, resides with the observer and his observations or with the universe and its forms is indifferent. Risk remains endosystemic, because future can only be endosystemic.

But future is ignorance. Future is what the system does not know. Future is a locus of unutterance, the continuum of origin, the rupture of the space before difference.[36] Everything the system ignores comes from the future – the system's future – considering that everything the system knows is in the present and whatever it may learn will be lying in its future. But the future lies within. In projecting the present, the system incorporates its future present and attempts to weigh, not uncertainty which is granted, but its reaction to the measurements of profits and losses that come out of the projection. But to do this, the system marks and crosses sides constantly, thereby actuating contingency. The system incorporates its environment (future presents) in its boundaries (past presents) to select it by projecting onto it its past, thereby deselecting it.[37] But its selection is never a selection proper, but always a *withdrawal* from selecting: the system can never mark its future. What it can do is expose itself to the limits of the future within, the risk and its vanishing act. Thus, whatever the system attempts to do with risk, it always takes place in the systemic movement back from the space of ignorance. The system must withdraw from exploring its risk, and this is the only risk comprehension that it can perform. This is not an entrance in the space of ignorance, but a withdrawal from it, but *only after* it has tried – and failed.

The fact that the system can only project within its own future means that any causality between systems must now be seen as simultaneity, 'routinely co-ordinated' in such connections as systemic couplings and interpenetration.[38]

---

36 Philippopoulos-Mihalopoulos, 2006. See also below, Chapter 5, on the concept of the future as always to come.

37 Of course, the environment can never be selected as environment, but only as the contingent other side, which means that a selected environment changes name and place with the deselected system. This can only be momentarily, an example of which we will see below with the precautionary principle.

38 Luhmann, 1993a:98. Cf. Baecker, 2001:63: '[t]he notion of a system gives up the idea of ordering the world causally by attributing causes and effects to its different phenomena. Instead, it proceeds from the assumption that there are always too many and too few causes and too many and too few effects to be taken into account.'

However, while simultaneity is the only level of communication amongst systems, it also problematises communicability, because it entails uncontrollability of events,[39] something that becomes especially obvious with the shift from structural couplings to couplings of environments. The levels of uncontrollability are being reinforced by the undomesticatable complexity invited by the system, which becomes particularly embarrassing when the future of various systems is tied together, just as it is in cases of ecological concerns. But simultaneity is the external side of causality, and as things go, it has an internal side too – this time not from Luhmann, but from Edgar Morin, who, following chaos theories,[40] attempts to reconcile the paradoxes of quantum theories with considerations of continuum between humans and nature.[41] In his seminal work *La Méthode*, Morin suggests a circular, recursive, 'self-generated/generating' (*auto-générée/générative*) causality depicted by a closed loop, where any stage is both initial and final, both cause and effect.[42] Describing causality in such an endogenous way does not upset the communication between systems – at least to the extent that such communication is not understood traditionally as contact or sharing or mutual understanding, but as a surprising happenstance, all the more improbable and, therefore, 'real'. Endocausality, as defined by Morin and combined with Luhmannian simultaneity, confirms the paradox that all risks are risks to the extent that they are in any way perceived by the system, which in its turn allows the inference of a systemically enclosed presence of systemic future as the only way of perceiving and 'controlling' decisions about future risks.

II THE FUTURE OF SYSTEMS

The internalisation of the systemic environment in the above described form of projection of one temporal region over the other echoes what Luhmann calls 'time binding'. Time binding is the process of projection of a form (i.e. both sides of a binarism) in the future, in an attempt to create a space of determinacy on which expectations are going to be based.[43] Luhmann describes risk as a form with which the system confronts the problems posed by future uncertainty.[44] Risk is a form for dealing with time. With the

---

39 See Luhmann, 1998a, and King, 2001.
40 Indicatively see the classic Gelick, 1988; Briggs and Peat, 1989; and for an application on cities, Batty and Longley, 1994.
41 Morin, 1977, 1986.
42 Morin, 1977:258; Morin also calls it *endo-causalité*, the inner causality that 'brings about the permanent transformation of states which are generally improbable, into states which are locally and temporarily probable' (1977:259).
43 Luhmann, 1995a.
44 Luhmann, 1993a:51 and 1998a:71.

projection of this form the system binds time, namely endows the probability of future events with a historicity that derives directly from the memory of the system.

In the urban system, risks are multiplied together with the multiplication of temporal regions. In cities and for cities, projection is not easy. Time in all three manifestations appears multiplied by the intensity of corporeal perception and the frequency of environmental irritations. Historicity of the past and contingency of the future concentrate in the plurality of the urban present and continuously select and deselect each other, multiplying alternative pasts and futures and ultimately multiplying the possibility of present projections: 'the air is thick with time' as Grange remarks in his urban cosmology.[45] Present acts of temporal meaninglessness are taken up by temporalised spatialities and rendered significant in the projection. Forms typically exaggerate themselves by posing not only as extremes but also as selections that transcend the systemic horizon.[46] Risks are riskier in the city. There, one can see risk in all its tube-like perplexity, a net of bifurcations that eats up its own tail: the risk of risk becomes more significant than the risk itself. The projection becomes a representation and the weighing becomes light-headed. The body is one factor of this multiplication, with all its multiple time zones and functions that, as Michel Serres writes,[47] operate both locally and globally, bringing space and time together to create new forms of makeshift safety and fleeting familiarity – what in the next chapter will be described as *place* – only to depart from it later. Through its corporeal/spatial meaning, the city projects and selects from endless forms of future risks: safety and violence, control and anarchy, autonomy and isolation, dullness and stimulation and so on, all of them echoing and breeding the basic binarism between the enticement and fear of the contingent. In the city and for the city, future is as broad, luminous and available as it is unsettling, exposing and ultimately catastrophological. The future of the city is described by what Grange, following Mead, calls *inscape* and *contrast*, namely the singularity of the actuality of every event in combination with the continuity of other potentialities – itself not far from the form continuum/rupture characterising the boundary of the city.[48] The city faces its future with combined urgency and hesitation, enticed and at the same time intimidated by its potentiality, indulging risks as manifold and fractal as the selves of the city, each one a micro-niche of risk fermentation, each one replicating what risks there are and all of them constructing the urban risk.[49]

45 Grange, 1999:37.
46 See Pile *et al.*, 1999:11, on the 'felt intensities' of the city.
47 Serres, 1982.
48 Grange, 1999:21–40.
49 On the specificity of urban risk see Batty and Longley, 1994.

The other factor, of course, is space – itself only schematically disconnected from the body. Spatial forms used to be enough for the city to project: simple binarisms, like inside/outside, were adequate to help a city take risks and expect profits rather than losses. Walls around the city assisted selections and controlled risks. Now, spatial problems have assumed a different perspective, one of internalisation rather than external threats. This is exacerbated with the ingress of ecological risks that add to the social and become internalised and internally multiplied. Projections are literally thrown out of the spatial bounds of the city and the risk is now whether the city will expand so much as to threaten the remnants of nature around the cities. In the same vein, the problem of urban hinterlands – or 'ecological footprint' as it has been called[50] – that constantly expand to cater for the consuming needs of the city, indicates the historical inversion of the role of the environment from threat to threatened: once again the threat comes from within.[51]

Environmental law is one of the systems with which the city constructs a simultaneous present and attempts to accommodate its projections and decisions.[52] The coupling between environmental law and the city is particularly interesting in view of what Luhmann writes about law and its abilities to deal with risk. Law operates with norms, and through these law binds time in the form of stabilisation of expectations.[53] Norms are presumed to be 'risk-free' structures, in that risk should appear exclusively in deviance from the norm: if one follows the norm, one will run into no risk.[54] The problem is that law is overburdened by what is required of it, namely 'to bring the future into the present'.[55] The law can determine how others ought to behave in the future on the basis of its time-binding operation. But in the case of risk 'we are not dealing with a future for which we can in our present determine how others are to behave in future situations. A risk cannot be violated. If the law can be expected to assume risks, this can only occur by detemporalizing the assessment of what is right or wrong.'[56] This, however, according to Luhmann, leads to the paradox of a legal decision which may be valid because some future consequences have been foreseen, and not because of the inherent validity of the decision regardless of unexpected developments.

The above is true as long as one respects the dominant description of law as 'time binder' and qualifies neither the duration of time it is supposed to bind

---

50 Massey, 1999; Hinchliffe, 1999.
51 Haughton and Hunter, 1994.
52 Other systems include politics and science, both of which will be discussed later in this chapter. Economy is another relevant system, which, however, has been adequately explored by Teubner and Febbrajo, 1986.
53 Luhmann, 2004, 1993a.
54 Luhmann, 1993a:55.
55 Luhmann, 1993a:60.
56 Luhmann, 1993a:59, emphasis omitted.

nor the margins that such a responsibility leaves to law for post-decisional self-correction. I reserve doubt on these issues for later, where decisional issues are discussed in relation to time. Here, I only want to suggest that law – and I will only talk about environmental law although other specialisations come to mind too[57] – does have a way of doing precisely what it is supposed not to be able to do, namely assume risks by projecting a form (which already contains a selection) to the future. As I explain in the following section, the precautionary principle as a legally binding principle of environmental law does exactly this. Overburdening of the system is avoided by a postponement of the paradox of non-selection, which takes the form of assuming risks without quite indulging them. Doubts on how successful this operation really is will also be dealt with in the following section. For now, however, it is apposite to say that in the coupling between city and environmental law, the precautionary principle decidedly curtails temptations, limiting selections and suggesting to the city a simplified version of environmental complexity, which although less alluring, at least has the benefit of the familiar.

III THE PRECAUTIONARY PRINCIPLE

If the Sirens of the epos are the mythical representation of risk – one amongst several in a string of decisions described in the Odyssey – and the ship the system before the risk, one could describe Odysseus's approach to risk as precautionary – precisely because he threw himself into risk by withdrawing from it. In environmental law, the precautionary principle posits that the absence of scientific certainty should not be used to take or omit to take action that may be harmful to the environment.[58] This entails an assessment of scientific (un)certainty and a consequent weighing of what is expected to what is available, together with an assessment of the potential environmental losses of acting or not as the case may be. This exposure to the risk, yet self-induced protection from it, enables the ship to float by the isle unscathed yet wiser. This is also the environmental legal formula that deals with risks in the face of ignorance.[59] Precaution is employed when the risk is high, 'so high in

---

57 Luhmann admits some reservations with regard to criminal law (1993a:60, n.16). Other examples include strict liability, and generally the use of presumption.
58 Principle 15 of the Rio Declaration on Environment and Development states that 'lack of full scientific certainty shall not be used as a reason for postponing cost-effective measures to prevent environmental degradation'.
59 The precautionary principle is not the only tool of risk perception and comprehension, but it is the only one that deals with precaution as opposed to risk assessment or prevention. Environmental Impact Assessment is a widely recognised legal tool that includes risk assessment: see at the European level the Directive 85/337 [1985] O.J. L.175/1450 on the assessment of the effects of certain public and private projects on the environment, and particularly its

fact that full scientific certainty should not be required prior to the taking of remedial action',[60] and regardless of whether a cause-effect relationship between an activity and its environmental effect has been established.[61] The origin of the principle can be traced back to the early 1970s when Germany decided on a paradigmatic change of regime[62] from 'a law of environmental *allocation*' to 'a law of environmental *protection*'.[63] It is clear that the principle contains a presumption in favour of environmental protection when there is scientific uncertainty and high risk. However, within the principle, a certain assessment of both the significance of scientific proof and that of the risk is included – in other words a *projection*. But the projection stops short of resulting in a decision on the basis of the data collected. Instead, a presumptive mechanism is set off which eliminates the need for an actual risk decision and replaces it with a pre-ordered selection.

The typical binarism of environmental law (lawful/unlawful) is applied as a form between, on the one hand, what is usually a stable, internal fact (environmental protection), and on the other, a variable, external fact originating in other systems, such as the political or the economic. Science has a significant role in the selection and the difference between coding and programming can be of help here to understand its role.[64] Even if it presents itself as an external fact to which a coding value should be applied (i.e., the profits of animal testing for the advancement of science), science – in the form of legal proof – is what Luhmann calls the *criterion* according to which

---

recent amendments – Council Directive 97/11 [1997] O.J. L.73/5 amending Directive 85/337 – which recognise a greater role for risk assessment. Liability also forms part of the comprehension of risk, in the sense of the post-risk remedy as shown in Teubner *et al.*, 1994. Closely associated to this is the Polluter Pays Principle, again a part of the later stage of risk comprehension. Risk assessment is described as a 'cumbersome approach requiring significant resources and administrative effort', which only 'depends on what is quantifiable', and which only reflects limits 'imposed by the interests of key participants rather than those of the environment'. On the other hand, the precautionary approach recognises that 'science will not provide clear policy prescriptions . . . Instead of attempting like risk assessment to reduce uncertainty through a systematic, quasi-scientific process it focuses on the policy process itself and seeks to extract maximum response from legal and economic structures.' (All quotes from Moltke, 1996:101.) Prevention, on the other hand, is harder to distinguish from precaution, with some authors arguing that there is little point in the exercise, since most legal texts contain a combination of the two (Hohmann, 1994:334).

60 Kiss, 1996:27. Another distinction is made by Hohmann, 1994, who notes that the division is particularly obvious in English law, where prevention is more about tackling pollution at source, whereas precaution deals with the uncertainty of polluting substances.
61 Nollkaemper, 1996:74; also O'Riordan and Cameron, 1994.
62 Hohmann, 1994:5–12; Boehmer-Christiansen, 1994.
63 Hohmann, 1994:11, original emphasis. This has been characterised by the same author, 1994:3, as a 'reformed or broader anthropocentric approach'.
64 See above, Chapter 1.

a programme is established.[65] A criterion guides a selection which links a coding value (lawful/unlawful) with an external fact – and this combination is what Luhmann calls programme. The role of science – along with other considerations – is always important as a guide for environmental legal selections. Specifically in an application of the legal code through the precautionary principle, however, science becomes paramount in that, while it begins with its usual role (scientific proof as a criterion), it soon becomes the *only* criterion for the selection in view of its inability to be the criterion for the selection! Thus, precaution entails a foray into science, only to discover that science cannot be used, which renders science immediately operable as the only criterion in the form of presumption in favour of linking 'environmental protection' with 'lawful'.

To put it in risk terms, science is employed, first as an insight for the *perception* of risk, and immediately after as a pedestal on which a presumption rests for the *comprehension* of risk. For, in the precautionary principle, the two stages of perception and comprehension of risk are interfolded by the specific projection: the decision maker, by applying the precautionary principle, projects a/the present into the future and *withdraws* the projected present before they even perceive it. The decision maker 'suspects' that the present they have in their hands is neither 'the', nor exactly 'a' present, it is neither known nor an alternative present that they can know. The present of the precautionary principle is one of a short, jolted perception and an immediate negation of comprehension: I project it, I do not understand it, therefore I do not project it *but only after* I have projected it. Needless to say that I must have already projected it in order to comprehend that I cannot understand it. Risk projection is located on the very withdrawal from the space on which the present is projected. The risk, not being comprehended in terms of effect but comprehended in terms of itself (risk), is not acted upon.[66]

Clearly, science as a criterion is present in both parts of risk definition, namely perception and comprehension. Science is both the reason of the non-comprehension (I cannot comprehend the effects) and the momentum of the perception (I project the present on the basis of scientific knowledge). I know that I cannot comprehend this risk, because of lack of scientific certainty; the fact that I *know* (that I cannot comprehend) is based on knowledge of the *limits* of science, or else, on the negation of science as a ground for projecting. Exactly because I cannot define the specific risk (perception), I have to avert it (comprehension minus perception). I perform a caducous projection, I withdraw it almost prematurely, like a jolt in the yo-yo that stops before the whole string has been unfolded, and I retrieve it to the safety of knowledge

---

65 Luhmann, 1989b:38–45. Note that science appears both as tool and as purpose, which is why it can be both a value and a criterion.
66 Better: it is acted upon by being negated.

(of ignorance). And so on *ad infinitum*, or until the future: until (and if ever) the scientific absence is replaced by scientific presence. Because only then can the postponement of decision (or 'the *safeguarding of ecological space* or margin for manoeuvre in the future' as O' Riordan calls the operation of the precautionary principle)[67] stop.

Two conclusions can be deduced from the above that have a bearing on the present discussion. The first is that the precautionary principle is not a deci-sion on risk (although itself a risky operation), but a presumption before the decision that postpones the decision until the selections are clearer. Through the principle, the system accepts its ignorance of the future and deals with it by not touching it. It tests itself on the internalised limits of its absent environment, and postpones the battle. The system has already embraced the absence of its future, namely the system's inability to know it. The maze of ignorance has to be brought into the system for its extraneousness to be registered as that. The inclusion of the absent environment is the locus of unutterance: the system institutes a space within which it speaks for no one and refers to nothing but its very absence.

The second issue is a response to Luhmann's dismissal of law as a risk-taking system.[68] The preceding discussion partly confirms and partly reverses such dismissal. The only way the legal system can assume risks is through its norms, themselves risk free. If a norm perpetuates how one ought to behave, the precautionary principle as a norm perpetuates, until further notice, what one cannot decide. The risk is internalised but alleviated by the principle's inherent postponement of decision. The principle does not avoid risk: it sim-ply transfers it into its legal structures and the risk becomes a legal risk: was the decision correct? (And this is where Luhmann is correct, as I discuss in the following section.) Only the future can tell, but for the time being the future of the system has been converted by the principle into a faithful copy of the system's present. The projection of the present is stopped short from touching the future and sculpts instead a present out of the future. The precautionary principle freezes the present into the future (or the future as present), but only until the future actuates itself as different from the present – that is, when scientific data become sufficiently conclusive. This way of assuming risks is inherent in anything that attempts to dress the future with the present and stubbornly ignore other presents. It is the way another environmental legal concept, that of intergenerational equity, assumes risks.[69] The concept determines the interests of future generations solely on and invariably to the interests of present generations, which are assumed to

---

67 O' Riordan and Cameron, 1994:18, my emphasis.
68 A response which is not too distant in substance (albeit quite distant in methodology) to Paterson, 2003.
69 World Commission on Environment and Development, 1987. See also Chapter 5, section III.

include a healthy and clean environment. The time-binding operation of the law is exemplified in the form of unimaginative conventions: if future generations want to live in a city built out of glass, then one's efforts to list historical buildings are eminently misplaced. But at least these conventions are not at variance with what we *bona fide* think that present generations wish for.

To return to the metaphor of the Sirens, the ship projects her present into an already determined future. The determination comes from the presumption against risk. The difference to the precautionary principle is that, in the epos, there is no time for postponement: epic time is the time of narration and narration has to assume risks when they appear. So, postponement is replaced by the self-imposed limitations of knowledge. The knowledge of the song can come only after the ship has been exposed to risk, but then it will be too late: the system will have succumbed to what is deemed a lethal risk. Knowledge has to be safely encased in its boundaries, so that it does not spill over its predetermined limits, however strong the temptation. In that sense, knowledge is used to limit knowledge, just as science is used to limit science. But the comparison between the epos and the principle stops here because of the incommensurability of the stories: in the epos we know what has happened and every time the narration confirms the conclusion of this adventure and the advent of the next. In the case of the precautionary principle, practice obscures conclusions. It is not the case, for example, that every time there is scientific uncertainty the precautionary principle is mandatory.[70] Mitigations of the principle come from various sources, including the principle's own national[71] and international[72] versions in legal texts. On the

---

70 Nollkaemper, 1996.
71 In English law the concept is mitigated by two notions, both, according to Warner, 1994, rooted in the principle of precaution: first, specifically for setting thresholds, the ALARP ('As Low As Reasonably Practicable') which refers to technological factors such as quality control in manufacturing, maintenance, post emergency procedures etc; second, the BAT-NEEC ('Best Available Technology Not Entailing Excessive Costs') which introduces economic weighing. In the US National Environmental Protection Act, the requirement of 'worse case' analysis, where in view of scientific uncertainty the worst possible scenario should be taken into account, has been recently replaced by the 'rule of reason', which refers to 'reasonably' foreseeable events based on 'credible' scientific evidence (Shelton, 1996).
72 Although most of the treaties do not associate the principle with economic considerations (see Nollkaemper, 1996:76), the use of qualifying adjectives is widespread: e.g., both Art. 3(3) of the United Nations Framework Convention on Climate Change, Rio de Janeiro, 1992, and Principle 15 of the Rio Declaration on Environment and Development, Rio de Janeiro, 1992, where precaution applies only to 'threats of *serious or irreversible* damage'; Art. 2(2)(b) of the Convention for the Protection of the Marine Environment of the North-East Atlantic, Paris, 1992 where '*reasonable* grounds for concern that pollution may be caused' are required; or even Art. 2(5)(a) of the Convention on the Protection and Use of Transboundary Watercourses and International Lakes, Helsinki 1992, where the states are 'to be guided' by the principle, rather than apply it. For the justiciability of the principle from an international perspective, see Fisher, 2001; for a general and contextual overview, see Feintuck, 2005.

international level specifically, the precise status of the principle is disputed: some argue that it is not an established principle yet,[73] while others seem to be at ease with its generality and margin for interpretation.[74] Despite the inherent limitations, it is acknowledged that the precautionary principle is 'far more than a plea to add an extra margin on safety to old practices. It is based upon the realisation that it is extremely difficult to determine "safe" levels of contamination.'[75] Indeed, the main thrust of the precautionary principle has been exactly this legislative desire to anticipate scientific uncertainty by institutionalising a means of 'boxing' uncertainty, without however (and this is where realism in the form of postponement or epic ambition in the form of knowledge comes in) resorting to an absolute freeze on progress – which would have happened if the system were to use it simply as its external reference, thereby domesticating it. In view of this flexibility, the operability of the principle has often been exemplified in domestic legislation, such as the example of the §109(b) of the US Clean Air Act, that allows for national air quality standards to be set without requiring or even permitting the competent authority to consider costs and technology.[76] On a regional level, the European Union's decision to ban genetically modified organisms until more was known of their effects is an example of how postponement operates.[77] On an international level, a good example is the moratorium on large-scale drift net fishing on the high seas, imposed by the UN Resolution 44/225, which addresses the matter despite surrounding uncertainty, while also reversing the onus of proof.[78] Successes and failures alike, however, concentrate in the decision whether the precautionary principle can be applied in a specific case or not. Decision makers, in the formal guise of law makers and judges, and in the informal guise of political parties, experts, non-governmental organisations, and the general public share the problematic of administering their subjectivity, which, in its turn, is channelled to the risk of taking a decision on whether there is risk or not. Such a decision cannot be postponed and its paradox has to be solved by selection. This can only mean that we are back to the criterion of science, which is also redefined in view of its own questions of applicability, namely how much science is not adequate. Thus, the question

73 Birnie, 2002; Bodansky, 1991.
74 Cameron and Abouchar, 1991; Nollkaemper, 1996.
75 MacGarvin, 1994:70; see also Cordonier Segger *et al.*, 2003, for an optimistic reading of the principle following Johannesburg 2002.
76 Nollkaemper, 1996:79, n.27.
77 EC Directive 90/219/EEC, even if in its rationale protectionism vies with environmental considerations. For another European community example of risk and the precautionary principle, see T–13/99 *Pfizer Animal Health SA v Council* [2002] ECR II–3305.
78 Freestone and Hey, 1996:260. However, the moratorium is not as absolute as the Clean Air Act example, because of terms such as 'unacceptable impact of such fishing' and 'sound analysis'. The difference is understandable in view of the increased difficulties of international agreements.

departs from the confines of the legal system or, indeed, remains exposed to potential structural modifications as determined by the structural coupling of the environmental legal system with science, politics and public participation.

IV SCIENCE AND POLITICS

Risk's peculiarity as the cardinal embodiment of future uncertainty exposes an inadequacy of internalisation. While the latter remains the only way in which the system can perceive and comprehend systemic risks, the decision on whether these risks are risks and how to deal with them is constantly 'dealt to', passed over to other systems. This does not mean that the system does not face its risks to the extent that it can. It simply denotes a systemic incapacity to assume risks conclusively in the form of a decision that mutes future risks at present. Thus, while risks are dealt with within the system, other risks open up in a chain of risks in all systems that encompass both perception and comprehension of risk (since comprehension necessarily harbours further risks), and refer either to the decision on whether something is a risk or to the decision on whether to assume a risk or not.[79] The two differ in their temporalities: past or future of the decision-making system may be future or past of another system that may well have the answer to the question, but is prevented from imparting it, because of the temporal and causal incommunicability. In other words, the risks of risk extend in past and future of other systems and the chain bifurcates both backwards and forwards.

Ecological risks are a good example of the multiple appearance of risk. Law relies on its internalisation of scientific findings to actuate its decision-making operations; science, however, is often presented by the political system in political binarisms that are determined by votes; this alters the way science is internalised by law, especially when economic versus environmental weighing has to take place, which will also be presented by the political system and will resonate in economy as viable or not viable. This does not denote a shared risk in the manner of shared future, but internalisation of environmental irritations. Each system is responsible for its own interpretation of risk and deals with it the way it can. What the above does mean, however, is that ecological risks appear unique to each, yet common to all systems in a string of further and previous risks forming vinous extremities begetting other extremities in an ever-expanding chain-like discontinuity that

79 Luhmann, 1993a, uses the metaphor of a decisional tree to describe the same phenomenon. I prefer the idea of a chain because it describes better the autonomy of risk development regardless of decisions, and also the two-ended expansion in past and future. The same, but from the point of view of decision making, is eloquently put by Murphy, 1997:182: '. . . the process of deferral and displacement: things are decided elsewhere, and since things are "ultimately" always decided elsewhere, decisions are always being passed on'.

links past and future of all systems. Each system deals with the parts it can, but 'dealing' here can only mean postponement of the risk-to-come. In that sense, Luhmann's observation that systems like law and economy cannot assume risks is qualifiedly true: they can only assume their own risks and any decision on them is simply another etiolated shoot in the chain, always in search of the 'light' at the end of the tunnel, which is regrettably the knowledge of the past ('I have survived it') that somewhere at the end of the tunnel lies the light. But the end of the tunnel announces the end of time, which, for the time being, happily remains extra-systemic so systems can carry on producing and dealing with their risks.

The above is presented by Luhmann as the result of the combination of social and future uncertainty.[80] Not only future is uncertain, but social reactions to risk are uncertain too. And the way a risk is presented by the political system may be as important as, if not more so than, the descriptions of the experts: regardless of whether a nuclear power plant only explodes every 12,000,000 years, we do not want it near us.[81] Risks are open to individual cases, which are determined not only by future but also by social uncertainty. Thus, risks carry on being built up, burdening the system with decisions on the risk of risk. The urban system contextualises the chain effect clearly. In cities, risk of risk (schematically past) is as present as risk of no-risk (schematically present). The insulated urban environment produces its own structures that are only related to the urban processes as determined meaningfully within the urban space. This results both in exaggeration and in discounting of the perceived spatial isolation. In cities, the risk of risk brings fear which then breeds fear of fear, or else the annulment of risk decision, not as precaution, but as immobilisation.[82] The corporeal proximity imparts the impression of a common ecological fate, a shared future of ecological horror. This is what Luhmann calls overproduction of *resonance* of ecological issues within society. By resonance, Luhmann means the reverberating reaction of the system to environmental irritations, but always according to its structures.[83] For Luhmann, the problem with ecological issues from the point of view of society is not so much the potential catastrophes or the everyday

80 Luhmann, 1993a:48–9.
81 Interestingly, distrust of expertise is one of the conditions for the increase in demand for statistical data; see Murphy, 1997:137; see Sunstein's 2002, comprehensive comparison between risk perception and cost-benefit analysis.
82 The fear of fear immobilises the decision maker to a non-projection, because of the fear of the end of fear. Thus tells us the Chevalier in his dialogue with Blanche: 'Blanche: Do you think that I am kept here by fear? The Chevalier: Or by fear of fear. After all, this fear is not any nobler than any other fear. You have to learn how to run the risk of fear, as you run the risk of death.' From G. Bernanos's libretto to the opera *Les Dialogues des Carmélites* by Francis Poulenc. See also Bauman, 2005b.
83 Luhmann, 1989b:15.

environmental degradation, but the overproduction of ecological angst within society. In a city, ecological threats are internalised in situations such as technological accidents and intensified effects, because of the density of population, or through the tangible presence of various manifestations of the environmental movement.[84] The impact is intensified by the continuous presence of the media, and culminates in sensational totalising projections of an unsustainable urban present that manifests itself through social inequalities that can only get worse.[85] The individual case becomes uniquely common and the city faces the risk of disintegration of services, institutions, and representational capacity to the corresponding state or region.

At the other extreme, the urban system produces a risk of no-risk. The impression that there is no 'real' ecological risk finds a perfect breeding ground in the urban rupture with its environment, not only as recreational facility, but mainly as the commodification of products that are based on natural resources, but appear alienated from their initial forms: buying it in plastic packaging with colourful labels does not have the same effect as chopping it off an animal or collecting it in the field. The understanding of natural processes is replaced by an understanding of urban social processes – indeed, social processes are found in their natural continuum with natural processes. The city creates an allure of self-containment and nonchalance that bears little relation to its actual ecological impact. At the same time, it constructs a self-description of continuum with its social or natural extremes, while minimising its presence in the city's topology. The two reactions co-exist within a city, and the city as a system oscillates between token recycling and numbing fear, defining itself ecologically as the locus of social process threatened by the incomprehensible.

The system deals with its space of unutterance, and its risk more concretely, by concealing its ignorance: as Luhmann puts it, 'responsibility = the absorption of uncertainty'.[86] But how about the responsibility of not absorbing uncertainty? How about the exposition of ignorance in the form of uncertainty, the encounter with the limits in the form of oscillation, the presentification of absence as absence within the system? For better or worse, there are such operations, through which the system admits to the responsibility for its ignorance. One is by excusing uncertainty and heightening subjectivity; and the other, to pass on the risk, to deal it to another system, assuming, however the responsibility of dealing it to another system, of not dealing with it (at least not exhaustively, conclusively or convincingly). The former questions the possibility of scientific truth; the latter the possibility of

84 Luhmann, 1993a:125–9, where he states that they are simply a protest against functional differentiation.
85 Blowers and Pain, 1999.
86 Luhmann, 1998a:89.

democracy. In the former, risk is assumed as relativity; in the latter, risk is assumed not as uncertainty, but as responsibility by each system, only to be placed in circulation among systems as form where no selection has really been made: uncertainty absorbed by systemic rather than temporal postponement. The two can only schematically be separated since, on one hand, relativity tolerates subjectivity and, on the other, science is another system through which risk circulates. Arbitrarily, therefore, I start with science and continue with some remarks on the political system.

The precautionary dealing with risk expresses the impossibility of the Laplacian *démon*: the demon, by possessing an absolute knowledge of the variables that determine the state of a system in a specific time, is only a short step from predicting/knowing the state of the system in the next moment in time. The impossibility of knowing stems from the impossibility of expenditure: Lyotard famously parallels the expenditure of energy required to do such a calculation to Borges's story of the emperor who wished to have a map of his empire on a real scale: the project leads the country to ruin, because the whole population vests its energy in the task.[87] Another problem of course lies in the impossibility of knowing the exact current state of the system, because even if calculations are possible, small errors are always involved.[88] The final problem is the nature of the subject matter itself: there are systems that exhibit a non-deterministic behaviour which does not allow either prediction or reproduction, because of incalculable complexity which may arise from apparently simple events.[89] The above simply confirm the improbability of absolute truth. Luhmann solves this issue by subordinating truth to a specific combination of code (true/false) and programme (various theories).[90] Truth may exist within a code–programme combination, but its transposition is problematic, especially when other systems are involved and employ scientific theories as argumentation. In an enlightening passage, Luhmann notes the problematised relation between risk and science in view of absence of absolute truth: '[t]he risk situation of modern society has a double effect. To reduce risk, greater demands are made for scientifically guaranteed security within the context of probability/improbability, thus forcing them to adopt the rhetorical mode. Science itself may behave reticently, but in so doing it exposes itself to criticism that it does nothing to promote its comprehensibility, or to face up to its social responsibilities.'[91] Especially when economically informed science is translated into technology and then promptly introduced into everyday practice, risk becomes more informed in its subjectivity, in that everyone has an opinion,

87 Lyotard, 1984:55.
88 Cohen and Stewart, 1994:190 on deterministic systems.
89 Cohen and Stewart, 1994:22 on 'chaos'; also Lyotard, 1984:56, for the relation between uncertainty and accuracy.
90 Luhmann, 1989a:76.
91 Luhmann, 1993a:214.

often based on personal experience, on whether there is risk and whether the risk should be taken. Luhmann's solution to this is second-order observation: '[t]he observer of a decision maker may assess the risk of the decision differently from the decision maker himself; not least of all because he himself is not located in the decision-taking situation, is not exposed to the same pressure to decide, does not have to react as rapidly, and, above all, does not share in the advantages of the decision to the same degree as the decision maker himself.'[92] How 'objective' this can be, however, in view of autopoietic closure, second-order 'reality' (rather than reality), differentiated chronologies and endocausalities, is questionable. The consolation is that even at the level of second-order observation one cannot talk in terms of true/false, but only probable/improbable. This means that, while descriptions of connections between systems cannot be avoided in the face of risk, at least they are not presented as causalities, but *attributions*: 'it is now possible to observe how another observer makes attributions. . . . Attribution is thus itself seen as contingent, the attempt then being made to discover the factors correlating to types of attribution (personal traits, stratification, situational characteristics).'[93]

Attribution solves as many problems as it creates. One wonders, for example, whether attribution is a concession to social manipulation and coercion – concepts that do not normally appear in Luhmann's cosmology. Attribution appears dangerously close to what Tellegen and Wolsink have in mind when they write that 'a research tradition in risk perception developed, mainly because it was felt that the factors determining the public *misconceptions* about nuclear energy should be discovered and then *corrected*'.[94] This problem is tied together with another seeming concession to power-politics by Luhmann: the distinction between decision makers and those affected. In view of the allowed subjectivity following scientific relativity, one's natural reaction to the problem would be to look at democratic processes, especially public participation in environmental decision making for a solution to this division. Luhmann is quick to dispel high hopes. He rejects both the notion that all decisions have to be made in a participatory way, and that democracy is the domination of the people over the people, for the first one leads to decisions about decisions, and the second to an annulment of power by power.[95] In its ideal, democracy attempts the unachievable: to bring the parts into a whole by remaining faithful to the particularities of the parts. Democracy contemplates the big picture of society while not losing sight of the individual. And it was all well when democracy could be summed up as

92 Luhmann, 1993a:68.
93 Luhmann, 1993a:25–6.
94 Tellegen and Wolsink, 1998:166, my emphasis.
95 Luhmann, 1990a; instead, democracy should be understood as the binarism government/ opposition.

the dialectics of a relatively uniform urban society from which slaves, foreigners and women were excluded. Nowadays, this difference has been replaced by the difference between government and opposition, and, as a whole, democracy has emphasised the element of surprise of contemporary 'reality'. In Luhmann's words, 'I consider democracy an achievement that is evolutionarily improbable, full of presuppositions but politically realizable. This means, first of all, that one should not begin with the critique of its states and circumstances but should marvel that it functions at all and then ask how long it will continue.'[96] So, gratefully, one ought to wonder whether democracy, with all its differences and unities, can be analysed into something like consensus. But this has also been discredited from several sources as a tool for solving differences,[97] and most vehemently by Luhmann who, while hailing societal institutions (or else, 'the presumed opinions of unknown, anonymous third parties')[98] as a representation of an expectational hence fictional consensus, seems to be more prepared to accept the substitution of consensus by what he calls 'politics of understanding', which is defined as negotiated provisos that can be relied upon for a given time.[99] Politics of understanding is a method of fixing and postponing points of conflict so as to allow room for further coalitions and oppositions to be formed. The fact that they are strongly mediated by temporal considerations means that they have to be constantly renegotiated, thereby replacing the freezing effect of a truth-seeking consensus[100] with the pliability of the space of ignorance as to the most appropriate solution.[101]

This section could carry on in this mode *ad infinitum*. The frustrating oscillation between the various points of the risk chain, with their corresponding decision making, is dictated by the oscillation *perpetuum* between the epistemological 'realism' of autopoiesis and the visionary ability of a second-order observation. What this apparent paradox tells us is that the ability of each system to assume risks is co-crescent with the inevitability of decision transfer, either to the sphere of relativity or subjectivity. The former confirms the absence of totalising truth, whereas the latter attempts to bring together the Luhmannian distinction between decision makers and those affected, by admitting to a replacement of causality with an endogenous causality, much along the lines of attribution, but without a supra-subjectivist

---

96 Luhmann, 1990a:238.
97 'Consensus has become an outmoded and suspect value' (Lyotard, 1984:66); also Ladeur, 1995b; Teubner, 1989.
98 Luhmann, 1985a:50.
99 Luhmann, 1998a:69.
100 This is what happens according to Ladeur, 1995b, when Habermasian discourse theory gives priority to justice discourse.
101 This is why '[understandings] do not imply consensus, nor do they represent reasonable or even correct solutions to problems.' (Luhmann, 1998a:69).

presumption. The solution cannot easily come from democratic descriptions. Instead, one needs to turn to the specificity of each risk as perceived and comprehended by each system and attempt an enlargement of such specificity in order to be inclusive of the descriptions of other systems, which would only mean taking into consideration an absent environment within the system of observation. Hence the utility of the politics of understanding (or perhaps mis/understanding), which recognises the need to postpone in view of the impossibility to conclude.

Recognition of fluid functionality implies another recognition: that of pluralism or heteromorphy, in the form of second-order observation circuit,[102] as a decision-making factor. Lyotard's 'narrative knowledge'[103] as the knowledge of living, the one close to custom, oral transmission and everyday life – in short, the non-scientific knowledge, as the blind spot of the scientific method – 'legitimises itself in the pragmatics of its own transmission, without having recourse to argumentation and proof'.[104] This self-legitimisation is easily employed by other systems in order to obtain consent.[105] Even so, there are good reasons for which the separation between narrative and scientific knowledge cannot be strictly enforced, especially in view of the fact that scientists regard risk differently to the public: 'risk technicians calculate actual risks, the public is only perceiving risks.'[106] The difference between relativity of science, individual subjectivity and public opinion allow for a whole set of variegated factors of risk perception and comprehension, which suggest the need for an equally variegated theoretical tool to approach these divergent views. A tool which can encompass the priority of invitation of the various blind spots and spaces of the horizon that are to remain inaccessible – variably, those of subjectivism, not-yet scientific certainty, narrative knowledge, and so on. Douglas, for example, suggests cultural theory as a means of understanding individual projections: 'cultural theory brings us somewhat nearer to understanding risk perception of lay persons by providing a systematic view of the widest range of goals that the person is seeking to achieve. Instead of isolating the risk as a technical problem we should formulate it so as to include, however crudely, its moral and political implications.'[107] It could be argued that a cultural description of risk builds on the above described uniquely-common, chain-like risk as shared amongst systems: each risk has to be dealt with in its boundaries ('the question about risk has to be: how safe

---

102 Only in such a circuit, analogous to the one between environmental law and the city, do blind spots become visible in their invisibility. See Chapter 6, below.
103 Lyotard, 1984:18.
104 Lyotard, 1984:27.
105 Lyotard, 1984:28: 'the State's own credibility is based on that epic, which it uses to obtain the public consent its decision makers need'.
106 Tellegen and Wolsink, 1998:166; also Wynne, 1992.
107 Douglas, 1992:51.

is safe enough for this particular culture?')[108] without disregarding the circulation of risk amongst systems ('No mass media information, no consciousness of risk.').[109] Of course, the autopoietic suggestion cannot proceed from cultural (ultra-systemic) descriptions unless it is on the level of second-order observation as the contextualising force of distance (rupture) yet immersion (continuum). Even so, a second-order observer is expected to make choices and the whirlwind of public opinion as vehicle and instigator of risk is absorbing. Risk circulation rides and addresses public opinion circularly. In the political sphere, where one would reasonably expect pubic opinion to count, public opinion is described by Luhmann as a means through which political actors observe themselves and each other.[110] The internalisation of public opinion as a form of self-legitimisation relies on the employability of both narrative knowledge and scientific findings.[111] It all boils down not to hard facts of the kind risk analysis produces nor to what the public really fears, but to the internalisation of public influenciability through factors, such as 'representativeness' (if two power plants are located in the same area, and people are asked which one produces more noise annoyance, people tend to nominate the one which produces stench!), 'cognitive availability of relevant information' (information recalled about recent 'similar' cases, or cases that drew the attention of the media),[112] or less obscure factors (e.g., how the danger manifests itself, familiarity, trust towards the institution, etc.). But for such internalisation to work, both streams of knowledge (schematically: the interchangeability between observer/observed and blind spot) have to be employed. For the narrative knowledge, the sense of control that derives from the voluntary choice of risk[113] – from crossing the road, to smoking, driving, or having unprotected sex – is internalised and further offered to the relevant target group: a well-tested marketing tool.

But internalisation will always be lacking representability of either scientific prestige or public palatability, unless the distinction between narrative and scientific knowledge is abandoned before risks and the two proceed to the same side of the distinction, the other being risk. It has been argued that instead of receiving risk assessment and public opinion in two separate documents, the decision maker fuses the two by approaching the vernacular, informal knowledge, and by enabling it to integrate with or even direct scientific knowledge.[114] Because if, as the above author maintains, scientific knowledge is not a neutral, objective knowledge, but a form of cultural intervention

108 Douglas, 1992:41.
109 Beck, 1995:65.
110 Luhmann, 1989a:89.
111 Thus, Meyer, 2001; also Ferkiss, 1993.
112 Both examples from Kahneman *et al.*, 1982.
113 Luhmann, 1993a.
114 Wynne, 1996.

equally subjective as any layperson's knowledge, a much greater flexibility in what comes to be defined as lay and expert knowledge is required. This paradigm shift is what others have described as 'civic science',[115] 'consensual science',[116] or 'greening science'.[117] Whatever the term, they all express the impossibility of legitimisation of one form of knowledge without the other.

A practical example of fused knowledge in view of scientific uncertainty comes from the role of public concern as a material consideration for planning permission. Stanley draws on two cases where local public reaction to waste installations was deemed a material consideration to be taken into account on the question of costs by the court, even though the risk had not been objectively proven.[118] Narrative knowledge, according to Lyotard, needs no proof other than its transmission. As Stanley maintains, 'the public assess exposure to development-related hazards on the "evidence" provided by the media. Public concern does not therefore require other evidence to establish it as a valid material consideration. Other evidence is relevant only as to *weighing*, not the *legality* of public concern.'[119] Cases of scientific uncertainty, however, are easier to solve than genuine clashes between scientific data and public assessment. In such cases, more often than not, the legal system passes the risk onto politics. According to Luhmann, the quest for a 'rational' solution to the problem of divergent subjectivisms makes jurists transfer the question altogether from law to politics. Politics becomes the third value of the legal binarism, where everything that cannot be decided legally is passed over to politics to be decided politically. However, 'in this way the legal system makes use of a constitution and democratic legitimization to deceive itself that politics can handle problems better than law and that all arbitrariness can be transferred into this system for appropriate treatment and reintroduced as a legal norm. This, however ... leads the political system, on one hand, to view law as its own instrument of implementation, and on the other to decisions within the legal system that are not decided in a specifically legal way but are determined arbitrarily.'[120] The added problem with politics especially with regard to ecological risks is a difference in temporal emphasis: while politics operates with short-term programmes, ecological issues require long-term, patient programmes that do not aim for votes but for slow-moving, unassuming changes.[121] Still, the rare but hopeful occasions when an

115 O'Riordan and Cameron, 1994:66.
116 Lovelock, 1994:122.
117 Shelton, 1996:224.
118 *Gateshead Metropolitan BC v. Secretary of the State for the Environment* [1995] Env.L.R. 37 and *Newport County BC v. Secretary of the State for Wales and another* [1998] J.P.L. 377 in Stanley, 1998.
119 Stanley, 1998:993, my emphasis.
120 Luhmann, 1989a:73.
121 Luhmann, 1989a:91.

ecological risk is internalised and dealt with by the legal system shows that the system has the forms to assume risks, and through the legitimisation of public assessment of risk, the distance between decision makers and affected can be shortened. The affected stops being, as Luhmann wants it, 'an amorphous mass that cannot be given form'[122] and becomes represented within the legal system through its self-authorised presence.

V IN LIMITATION

The boat floats on her own reflection. The risk always comes from inside the boat, a looming fear of relentless desire to know and at the same time to abandon oneself. The Sirens become the individual ambition, the idea of the universe embottled in one's perishable body. The crew perceives the threat of the song (the archetype of intra-systemic risk), exactly because 'the strain of holding the I together adheres to the I in all stages; and the temptation to lose it has always been there with the blind determination to maintain it'.[123] Odysseus appears as a typically disinvolved decision maker who chooses to disregard the desires of his crew and deny them the possibility of risk projection. While this is not 'democratic' decision making, and certainly echoes Luhmann's 'not everyone can participate in all decisions',[124] Odysseus remains an epic hero, and in the society of risk, no heroes can be accepted.[125] Risk outside the epos is a mostly thankless everyday crawl in a chain of bifurcations where any choice entails and presupposes further risks with the lure of further knowledge. It is rare for knowledge to be offered without risk and it is impossible to decide on a risk without employing knowledge. Of course, the decisions on the kind of knowledge employed and the weighing of the knowledge – themselves not less risky – typically rest with the decision maker (whoever/whatever this may be). Integration of the desire of the crew in the decision-making process may have worked, if properly presented. But the epos is beyond the need for legitimisation. The only requirement is for the story to go on, for the ship to carry on sailing. In some respects, the same is required of a system: to carry on being. But in order to be and become, a system takes risks – its very own risks – by internalising the space of ignorance, the future as a future present along with an infinite amount of other future presents. The future of the system lies inside the system and languidly spreads its ignorance, offering its dangers to the system. And in encysting them, the system converts them into risks; but some risks cannot be dealt

122 Luhmann, 1993a:110.
123 Adorno and Horkheimer, 1997:33.
124 Luhmann, 1993a:105.
125 Luhmann, 1993a:103.

with, except by allowing them to remain present in their absence. The system projects and then withdraws, crosses slices of contingency without touching on anything but itself: it flies above its environment as it would fly above the nocturnal countryside. Occasionally it is given the opportunity to have a look at the darkness: a sheath of light, enough to alert it to the risk that may be lying ahead. In those cases, subterfuges of presumptions and postponements are evoked, feeble attempts to stop or delay the nocturnal flight. What is remarkable is not whether they can delay it for long, but that they denote a systemic awareness of limits, which points to an ultimate acknowledgement of ignorance. And all this, within and through the system's operations.

For just as the origin of risk lies within the system, the operations of dealing with risk form part of the operations of the system to which the particular risk has been allocated. The boat, a languid system of crew and captain, sails by the formidable risk unscathed and victorious, but not in ignorance: the ears of the system are open to the risk. The disinvolvement of Odysseus has only been impressionistic. The risk has been defeated through its very chosen bodies of appearance; the boat, sealed against the risk, manages to accommodate the influx of knowledge: the system, in its secure closure, manages to accommodate ignorance through its openness. Risk commands for ears sealed with wax *and* ears open to the call of knowledge: for it is equally unrealistic to seal one's ears to knowledge, as it is to keep on exposing oneself to the Sirens.

Chapter 5

Community

Withdrawal, intergenerational equity and environmental rights

Community and rights are intimately linked with both environmental legal and urban considerations. Perhaps the most significant locus of convergence is the representability of either with regard to the environment. Thus, both legal and ecological community debates centre around the community's ability to speak for the unspeakable, while rights are often seen as one of the vehicles for such representation. This is indeed the beginning of this chapter's debate, whose adequacy for the kind of considerations put forth in this book will, however, be thoroughly disputed. While recording the importance of ecological and humanistic nostalgia towards community, this chapter attempts to construct a notion of community that does not fit in what seemingly is the panacea for our ecological woes.

In doing this, the text performs a very explicit withdrawal – indeed, a withdrawal from Luhmann's writings. This is a departure tout court*: from Luhmann's preference not to deal with community, as well as from Luhmann's poetics of language. This is not coincidental: it is not possible to talk about the absence of community in autopoiesis, mainly because, to some extent, autopoiesis is predicated on the absence of community. Any observation of community would depart from the Luhmannian focus on the social. Hence, autopoietic terminology will have to be enriched in order to shift its epistemological stare towards these excluded territories, which refer, not to the internalised and domesticated absence of community with which each system deals, but to the kind of community with which the system is to remain unfamiliar. This chapter, therefore, departs from Luhmann's opinions and language, but not, I would like to think, from autopoiesis as such. In fact, the withdrawal is here presented as an operation of autopoiesis in more ways than one. Thus, the withdrawal is itself conceptualised as the location of community. In its departure from nostalgia and its couching in social rather than conscious terms, community – or rather, its absence – becomes autopoietically operationalised. Following the discussion on risk, community is discussed here as another example of the continuum/rupture form, especially in its location on the very movement between outside and inside. At the same time, community develops the form in ways risk does not. Thus, community is connected to the messianic, which, however, does not arrive in the guise of future risk, but as a fractal reiteration of the form in the folds of the present. The connection of community with the environmental concept of intergenerational equity occurs on the basis of a present,* transient *recursivity, thus semiologising community as absence.*

This absence is further employed in the domain of environmental rights. The paradoxicality of rights is explored and a locus standi *of absence is adumbrated, in resonance of unutterance and withdrawal from collectivity. In search of the location of environmental rights, the discussion returns to autopoiesis with a conceptualisation of community as the circularity of observation on the basis of blind spot. Community is connected to rights, not through the collective, but through the absence of the collective as a representational entity that 'speaks' for nothing except its withdrawal.*

I COMMUNITY NOSTALGIA

Community discourses have a strong presence in the ecological movement because of the nostalgic ideal of small, localised communities that can operate in harmony with nature. These approaches have been questioned, not only on the basis of their feasibility, but more importantly on account of their inability to deal with issues such as segregation, exclusion of difference, and universalisation of values, especially with regard to contemporary urban environments. The ecological discussion is, of course, nothing but a fractal reiteration of the general discussion on community. Traditionally considered, community is an intermediate ground between the monad and the collective. Community purportedly brings forth the needs and interests of an individual and links them to similar interests of other individuals in order to project these interests convincingly and legitimately, or simply to bring individuals together and satisfy their need for sociality.[1] More recently, thinkers have disassociated community from identity, and have started looking in difference for ways in which collectivity can be expressed.[2] On a more philosophical level, a sizeable portion of more radical texts attempt to establish places in between the theses of identity and difference, which accept the being-in-community while retaining the singularity-of-being.[3] It is in the latter that I seek inspiration for a formation of a community in an environment of ignorance. But, before that, some general observations on the present state of community are of relevance.

Communities are traditionally conceived as face-to-face, intersubjective, local collectivities that reinforce and preserve identity structures against a sweeping externality – be this the city, the state, globalisation or god. Community is often hailed as the desired solidity against an identity-diffusing status quo – thus Benhabib's proposed sole avenue of social feminism: 'a solidaristic community that sustains one's identity through mutual

1 Communitarians emphasise collectivity, often in a conservative and nostalgic way, as the way out of the liberal individualism of capitalism. See Rawls, 1971, as the source of inspiration, and then MacIntyre, 1985; Sandel, 1982; Taylor, 1991.
2 E.g., Young, 2000.
3 Nancy, 1991; Agamben, 1993; Blanchot, 1988.

recognition.'[4] Often the community is a historical construction (invested with not inconsiderable amounts of nostalgia)[5] that erects its importance on the imagined sovereignty of its central tenets:[6] for there is always imagination, sovereignty and centrality in community idea(l)s. The members imagine their communion around their membership, thus delimiting themselves and the emergent quality of totality around such ideas as freedom of choice (to be in or out), sense of self-completeness, consensus, rationality.[7] Community is often seen as the bastion of civic (in the sense of political and urban) democracy, where decisions can still be taken by the ones affected. After all, community is a tangible manifestation of communist ideals in its flexibility of appropriation by capitalism, 'socialism' and liberalism. Thus, it can be employed in the services of diffusion or adoption of responsibility; as a support for and a counter-argument to integration; as an anathema or blessing for the dismantling of the welfare state; as a locality that resists globalisation or as a globality that resists parochialism: 'the elasticity of community is its appeal . . . [it is] offered as an object of appeasement or incitement, aspiration or accomplishment.'[8] Practice or goal, community is invariably addressed in a nostalgic way that reproduces the division, popular in the nineteenth century, between *Gemeinschaft* and *Gesellschaft*, namely the passage from intimate community to alienated society,[9] which, according to Jean Luc Nancy, is credited to Rousseau.[10] Rousseau was the first to reveal the connection between the two by disconnecting them, that is by experiencing society as the uneasiness towards community. And so, community had to be recreated as the ruins on which to build society. The kind of community needed in order to proceed to society was a 'much more extensive communication than that of a mere social bond (a communication with the gods, the cosmos, animals, the dead, the unknown)', namely a community that was not community in the present sense, but 'something for which we have no name or concept':[11] pure nostalgia for the return but perhaps not for the destination.

It is very difficult to sustain and justify such a sense of nostalgia, even for justified environmental purposes. The habitual encounter with a resuscitated locality, spatially concrete, environmentally connected, and with freedom of decision making as well as self-containment aspirations,[12] is inadequate (even

4 Benhabib, 1996:38.
5 Kellogg, 2005.
6 Anderson, 1991.
7 Sennett, 1990; Elster, 1998; Bohman and Rehg, 1997.
8 Singer, 1991:125.
9 See Tönnies, 1963; Luke, 1996.
10 Nancy, 2000.
11 Nancy, 1991:11.
12 See the relevant ecological communitarianism, e.g., Bookchin, 1990a; Eckersley, 1992; and Sandilands, 1998:251, who, from an ecofeminist point of view, supports localised democracy processes, on the basis that 'nature is specified locally'.

for utopian purposes) a tool with which to deal with the change of conditions, let alone with the conditions of change. Space is redefined beyond the confines of definitional certainty: borders are now observed independently of their inner or outer sides; propinquity is no longer a requisite for intimacy; striation is never independent from smoothness; private is increasingly controlled in a comparable way to public; a vocabulary of flow is now part of classic contemporary urban descriptions.[13] This does not mean that there is no 'order' in the way space flows – just that traditional descriptions of order are to be revisited.[14] In its asymmetrical equidistance from law and justice, 'order' defines the place of identity in an urban space of mobility: thus, one reads about hybrid networks,[15] liquid loves,[16] cyborgs,[17] atonal ensembles,[18] 'ecotone',[19] or indeed the 'authentic We', which combines a common project with individuality,[20] and one understands that even resistance to such apocalypses without apocalypse is part of an order that is being rapidly redefined in trans-spatial, trans-temporal, trans-human, trans-being conceptual regions. There have been calls for the disassociation of community from its fixation on human groups,[21] from locality,[22] or indeed from definition.[23] Such calls cannot be safely ignored.

Arguably, one of the most prominent problems of traditional definitions of community in relation to law has been eloquently put by Reza Banakar: 'the spirit of community underpinning community law is often the product of tightly integrated social groups. Such social groups as a rule *demand* a high degree of conformity from their members, as a result of which their community law might be turned into a repressive instrument of social control.'[24] If repression is the problem within, exclusion is the problem from without. Thus, Nicola Lacey: 'discourses of community may be thought of as speaking to the longing for a primal unity and connection; yet, much like the liberalist universalist signifiers which they in part purport to replace, their underside is always a process of differentiation and exclusion. The community which promises to connect and include is itself defined in oppositional

13 Amin and Thrift, 2002; May and Thrift, 2001.
14 Shields, 1997.
15 Latour, 1993.
16 Bauman, 2003a.
17 Haraway, 2004.
18 Edward Said, quoted in Chambers, 1994; see also de Sousa Santos, 1995; Taylor, 1989; Pile and Thrift, 1995; Bookchin, 1990a.
19 An ecologically informed paradox accommodating individual and social desiderata; Krall, 1994.
20 Vogel, 1994.
21 Knorr-Cetina, 1999; Latour, 2004a.
22 Gilroy, 2000.
23 See Agamben's *whatever* community, 1993; also Lasch, 1991.
24 Banakar, 2003:222.

relation to an Other who is excluded from membership and whose existence is denied.'[25] The exclusion of difference has famously been criticised by Iris Marion Young: attempts at urban community integration are seen as the main source of racism and chauvinism, as well as the kind of segregation that characterises most contemporary cities.[26] Community formation has been accused of encouraging members to assume the 'sameness' of approach of the focus issue (the *raison-d'-être* of the specific community), which immediately implies that outside community lies 'otherness'.

On a more abstract level, communities fail to proffer an alternative to what they frequently purport to replace: the centre. Such dictatorial thematisation is the consequence of a seemingly indispensable internal universalisation of community values, both as values that encapsulate communitarian principles, such as solidarity, equality, reciprocity, natural harmony etc., and as the values that constitute the interest-core of the particular community.[27] In other words, the criteria of belonging are a reproduction of the broader social rationale in which the community is located, and the communitarian universal is not far from the universal propagated by the surrounding status quo.[28] This reproduction is evident even in cases when one would least expect to find it, such as informal community justice, where structures of formal state justice are perpetuated within the informality of marginal justice,[29] or even in the case of environmental groups, which, in their politics of belonging, reproduce state patterns.[30] It is interesting, therefore, to observe here too the continuum/rupture linking inside and outside – here, a community and its environment. The fear of being one with the outside constructs safe effigies of environmental ignorance and internalises it as endosystemic differences. This is the crux of the matter in terms of community nostalgia: community is the home we've never had, but we always thought we did. Community institutes a pre-utopian space, before difference has become visible, before even a space has been carved from which to observe difference. In its self-contained limits, community includes continuum and rupture, and recreates the womb to which everything returns, the Eden which protects us from difference: 'what that word evokes is everything we miss and what we lack to be secure, confident and trusting . . . "Community" is nowadays another name for paradise lost – but one to which we dearly hope to return.'[31] But this in itself is a good reason to be suspicious – as Nancy puts it, 'suspicious of the retrospective

25 Lacey, 1998:134–5.
26 Young, 1990a; see also Sennett, 1976.
27 Young, 1990a.
28 E.g., Lacey, 1998 for a feminist critique; Fincher and Jacobs, 1998 for an urban-focused analysis, and Harvey, 1996a for a marxist critique.
29 As shown by Fitzpatrick, 1992.
30 Hannigan, 1995.
31 Bauman, 2001:3.

consciousness of the lost community and its identity (whether this consciousness conceives of itself as effectively retrospective or whether, disregarding the realities of the past, it constructs images of this past for the sake of an ideal or a prospective vision)'.[32] One is afraid without the nostalgia of a future community, itself operating as limits against ignorance and incentives to abandon oneself into the manic desire for a 'home'. Thus Derrida: 'Man *calls himself* man only by drawing limits excluding his other from the play of supplementarity: the purity of nature, of animality, primitivism, childhood, madness, divinity. The approach to these limits is at once feared as a threat of death, and desired as access to a life without *différance*.'[33] Life without *différance* would be life without uncertainty, without erasure, with solid instead of liquid, with ground instead of quicksand, with totality instead of continuum/rupture. The kind of life that one would need to leave behind.

Luhmann's departure from nostalgic *Gemeinschaft* considerations is absolute. He discounts the possibility of intersubjectivity, especially the Husserlian kind, because it is based on the monadology of consciousness.[34] Intersubjectivity for Luhmann can only be relevant if it produces social meaning, that is, if it departs from a self-restrictive humanism.[35] Of course, there are ways in which humans participate in society – interpenetration is one of them; socialisation is another.[36] But, as he points out, 'participation in a social system requires human beings to make their own contributions, and it leads to human beings' distinguishing themselves from one another and behaving exclusively for one another; because they must produce their own contributions themselves, they must motivate themselves'.[37] This peculiar last phrase points to what Luhmann sees as the basis of a community, namely the atomistic contribution to society; what is more, this contribution must be distinguishable from other contributions – in other words, the interpenetration must be reciprocal between the social and the conscious system. In addition to this, a schema of consensus/dissent operates as a constant comparison to what others do, therefore shifting the emphasis from the problem of intersubjectivity to the supposed solution of societal mediation between subjectivities.[38] Luhmann's fixation with *Gesellschaft* has been reinstituted despite that brief journey into subjectivity.

32 Nancy, 1991:10.
33 Derrida, 1976a:244, original emphasis.
34 Husserl, 1983; see also Derrida's, 1973, criticism, and Merleau-Ponty's, 1962, further emphasis on the body.
35 Luhmann, 1995a:80*ff*.
36 See Chapter 2.
37 Luhmann, 1995a:220.
38 For an analysis of Luhmann's socialisation as inclusive of the binarism consensus/dissent, see Vanderstraeten, 2000.

I will be showing later, through an insidious connection with Husserl, how Luhmannian observation can be employed in order to describe a community. Before that, however, there is another community worth exploring: the eco-logical community. In ecological formations of community, the nostalgia often appears as a moral call to overcome the 'biggest political boundary', that is the boundary separating the human from the non-human.[39] Deep ecological constructions of community entail a normative interdependence between human and natural, with ensuing resemiologisations of human interaction:[40] as Aldo Leopold, arguably the pioneer of ecological ethics, says, 'all interlocked in one humming community of cooperations and com-petitions, one biota'.[41] The self within the biotic community, according to the founder of Deep Ecology, Arne Naess, has as its ultimate goal the 'Self-realisation', namely the identification of the individual self with the total universe, which involves a Hegelian implosion of the concept of community.[42] Often, ecological ethics originate in technical ecosystemic studies, which are then analysed in their transferable ethics between human and natural.[43] Thus, states Holmes Rolston III in his influential ecosystemic ethics: 'The ecologist finds that ecosystems *objectively* are *satisfactory communities* ... and the critical ethicist finds (in a *subjective* judgement matching the *objective* pro-cess) that such ecosystems are imposing and satisfactory communities to which to attach duty.'[44]

The nostalgia for a collectivity that has never been, and a continuum with-out rupture, is readily embraced by environmental law as the panacea of the collective against the problem of the individual. The miasmic penumbra of property rights is apparently diluted only with the advent of collectivity, whose consciousness includes not only itself but also its surrounding possibil-ity of ownership.[45] Thus, through an unsurprising (but supposedly pragmatic) internalisation of ecological responsibility within the confines of individual interest, collectivity is reassuringly instituted as the means by which national, regional and international problems are to be solved.[46] Issues such as climate change,[47] global commons,[48] biodiversity, and so on, can only be sufficiently addressed via a 're'-stitution of collectivity under the *arkhé* of the common goal. The same issues appear on an individual level, as an escape from the

---

39 Dryzek, 1998:592.
40 See generally Devall and Sessions, 1985.
41 Leopold, 1953:148.
42 Naess, 1989.
43 Ellen, 1982; Caldwell, 1990.
44 Rolston III, 1988:167, original emphasis.
45 Coyle and Morrow, 2004, expose the problem without being critical of its fundamental problematic.
46 Hanemann, 1988; Campiglio *et al.*, 1994; Panjabi, 1992.
47 Zamagni, 1994.
48 Hardin, 1992.

latter and a return to collective action in the form of collective human rights which can represent environmental requirements more fully. Legal ecological communities are often seen as the only way in which the environment can effectively be represented in litigation.[49]

At the political end of the spectrum, an attempt to pragmatic collectivity returns with Bruno Latour. In a world where the concept of nature is abandoned, the enigma of the association of non-humans remains intact, and so does the possibility of collectivity between humans and non-humans.[50] Still, it is this very uncertainty that makes the specific community between humans and non-humans a space of debate. Latour believes in discursive rationality, but with embedded uncertainty in the form of speech impedimenta.[51] But it is in the formation of community on which I want to focus: Latour's collective is an internalisation of the multiplicity of either side – one side being the *polis* which involves cultural representation, and the other side is *physis*, which needs no representation since it provides reality. But this reality is indeed unrepresentable: it enters the collective 'in the form of new entities with uncertain boundaries, entities that hesitate, quake and induce perplexity'.[52] The environment is endowed with surprises and events ('no one knows what an environment can do . . .'),[53] and as such remains: its internalisation does not entail representation. On the contrary, the surprise is to be felt in the collective: this is not 'fabulous, dialectical, new, exotic, baroque, Oriental or profound. No, its banality is its best quality.'[54] This version of the collective, which draws heavily on environmental ethics, is designated to replace the dichotomy subject/object with human/non-human, thereby allegedly eliminating the issue of appropriation of nature, as well as the problem of centralised communitarianism. His collective is diffused, uncertain, unbounded. The community of humans and non-humans (Latour prefers the term 'association') is imagined as an operable state of uncertainty which accommodates the outside while silencing the subjective positionings of self-interest inside.

Although careful to avoid transcendental claims, Latour is flirting with the *au-delà* when he claims such appeasing qualities for the 'other side of the social'. His laboriously explained political programme is perfectly anti-transcendental, focusing, not unlike Luhmann, on the thing in hand, the societal tools of dealing with the problem of division between *polis* and *physis*. However, his utopianism is reminiscent of the ecological communitarian ideology of operability, of 'work', of delineation, of presence. There is

49 See the example of collective environmental rights, Anderson, 1996; Boyle and Anderson, 1996; Picolotti and Taillant, 2003.
50 Latour, 2004a:41*ff.*
51 See above, Chapter 3, section II on unutterance.
52 Latour, 2004a:75.
53 Latour, 2004a:80.
54 Latour, 2004a:79.

also a clear Hegelian happy ending when the exchange of democracy between the two kinds of associations results in a common (human) world. There is a self-realisation, a triumph of possibility, which is problematic in view of the ramifications of the idea and practice of community. In other words, there is a modest utopianism, dressed as a political programme. While this is not as such problematic, it may become so when one looks at the ramifications of such modesty. To put it differently, the above kinds of (ecological) community formation do not accommodate absence in its absence, but domesticate it, either by returning to the self, or by bridging objectivity with subjectivity, by transferring normativity between constructions, by reinstituting *logos* as speech and rationality (e.g., Latour's *speech imedimenta*), by containing uncertainty in a politically ambitious democratic project, by elevating collectivity as a moral value that transcends the typically individual by employing its self-interest structures, and ultimately by encouraging a continuum to prevail over rupture. The collectivity rationale, be it in ecological ethics, urban boundaries, environment law or ecological politics, suffers from a distinct lack of uncertainty – there is nothing less convincing than a self-assured formation of a community that thematises its totality in an inescapable way; a community which, however sensitive it may appear to its environment, cannot avoid presentifying it, rendering it readable and employable, bringing it forth as (not-just-a-)resource, as a (non-speaking) counterpart to a democratic project, and as a source of instructive ethics. Whatever the epithets, the nostalgia for collectivity consistently fails to decentre itself, even when self-doubted.

II ABSENT COMMUNITY

The nostalgia characterising and motivating the usual formulations of community brings forth an element of loss, of both having lost something and of being lost. The loss is experienced in a spectacular array of domains: politics, religion, philosophy, science, law. The sorrow of return (the etymology of 'nostalgia') is what pushes one to return as well as to resist returning. The construction of one's Ithaca is always utopian, in the sense of all-informing *and* never-reached: return (rather than Ithaca) is good for Odysseus, as well as for Penelope, for the text, and for the ones who listened to its recitation. The return, the change of direction, the other sense, is always an oscillation – not unlike Penelope's shroud, weaved during the day and undone during the night. In the unworking of the shroud, in the re-turn, the change of direction, Penelope wove the oscillation of her husband's return: Odysseus was leaving behind his community of the sea in order to be embraced by the community of the one. Odysseus is lost in an ambi-directional nostalgia, an oscillation between one in community and community of one (always left behind for Ithaca or for Iliad), the realisation that even if loss is retrieved, one remains lost. The oscillation between self and community is rarely left to its own pendulum – the paradox of being-with must be resolved, and an 'in-between' is the most fashionable response.

Veritable 'in-betweens' are a difficult affair. Especially with regard to community, whatever starts as in-between either reveals or hides a totality – nostalgia for the lost self or the lost other.[55] Loss appears in all its luminosity in the city, where nostalgia for the form translates into a melancholy of death. The city *as* death is the rupture of the completeness and its illusionary mending on the face of the other: 'the town, once beautiful and beloved too, embodied the loss he felt. Bruges was his dead wife. And his dead wife was Bruges', writes Georges Rodenbach in his elegiac novel *Bruges-la-Morte*, a search for the space between the city and the other's death.[56] The city loses itself in the palimpsest of miscommunication between the social and the conscious, and within the fragments of interpenetration, the city's descriptions are always 'at a loss', on the boundary and then knocked over, now inside now outside. The nostalgia of the form is the sorrow of return to the never-(to-be-)experienced: to a never-experienced Eden and a never-desired Gehenna. When Agamben revisits the Talmud, he brings forth the dual topology of nostalgia, itself always *double*. In both Eden and Gehenna, each person receives two spaces: one for themselves and one for a neighbour, respectively damned and saved.[57] For Agamben, this is the demise of the absolute of unsubstitutability, the ease of moving from oneself to the adjacent empty space, the unrepresentable community. These cities of absence, of emptiness to be populated by movement, of de Chirico-like enigmas, are what the city is: incompleteness and substitutability at once. In the city of detachment and involvement, where the city observes itself in a perpetual continuum/rupture with its environment, the need for simplification is overwhelming and the easiest field of simplification seems to be the community. The city proceeds to the stalls, reducing itself to a spectacle: 'the Greek city assembled in community at the theatre of its own myths.'[58] The city enters the secure and converts itself into a 'city of superlatives' as Beauregard calls it, where the conversion is absolute, sucking in its linguistic extravagance even seemingly critical observers.[59] The city is left to its own uncodifiability, and an urban signifier – the community – becomes the city. The effortlessness of such ventures is of course very quickly marred by what the community excludes, not least the city as the superlative absence.

How to stay away both from individual and collective, conscious and social, systemic and environmental, involvement and detachment, death of the city

55 See Fitzpatrick, 2001; Norrie, 2005, criticising Giddens.
56 Rodenbach, 2005:33.
57 Agamben, 1993.
58 Nancy, 2000:51.
59 Beauregard, 2003.

and death of the other, while acknowledging their position in a form whose completeness is never to be acknowledged? The process involves a meticulous description of the operation of boundary, which will then be linked to a process that trails the form place/space. Once these are analysed, the text can proceed to discuss the specificities of an absent community. This is not simply a quest for yet another 'in-between': this 'in-between', as suggested here, is not a third term, but the epistemological freezing of movement and directionality, but only in recursivity. What is proposed here is a conceptualisation of the in-between as movement, or more specifically as withdrawal, as a 'holding back' that allows for a space of ignorance to remain void.

Walter Benjamin has employed the term *porosity* in his descriptions of the city of Naples,[60] as the one linguistic gesture which could contain (by releasing) the noise, the bruhaha, the dusk of the South Italian city. Naples is the grand stockpot of merging dualities: new and old, public and private, churches and galleries, people and city, 'an unplanned, chaotic entity which constitutes what one might describe as an "organic" totality'.[61] Porosity describes 'a lack of clear boundaries between phenomena, a permeation of one thing by another'[62] by performatively allowing itself to take distance from the description, while pushing the observer further in. Porosity is 'the inexhaustible law of the life of this city, reappearing everywhere',[63] within and without the city, within and without the observer, the state of affairs of the very distance that links the two: the location of the city. This omnipresent law presupposes closure but also guarantees openness: it is on the closure that the pores are inscribed. The boundary is permeated with pulsating stomata of crossing. Porosity perilously treads on the cusp of bifurcation. A porous unit is closed but not isolated, identifiable but not impermeable. While not subscribing to the concept of an organised, constant unity, porosity advocates a certain closure extant in time and space while confounding its boundaries: 'the stamp of the definite is avoided. No situation appears intended for ever, no figure asserts it's "thus and not otherwise".'[64] The stomata of diffusion on the boundaries of the porous self erode its narcissism and replace it with a translucent crossing that occurs nowhere but on the very opposite side of its occurring.

The paradox can be landed on the withdrawal of place from space – what elsewhere I have called *placelessness*.[65] In the traditionally described form space/place,[66] the movement between the values of the form is epistemologic-

---

60 Benjamin, 1985.
61 Gilloch, 1996:25.
62 Gilloch, 1996:25.
63 Benjamin, 1985:171.
64 Benjamin, 1985:169–170.
65 Philippopoulos-Mihalopoulos, 2001.
66 E.g., Massey, 1994, 1995; Cooper, 1998. Here space is the presence of absence, while place is the security of presence.

ally 'frozen' and focused on the withdrawal of place from space. This movement (not unlike the movement encountered in the previous chapter) reinstitutes the absence of space as proper ignorance, unbound by knowledge. In the full movement, place remains simultaneously and non-simultaneously space, in that it can regress into form but also appear 'marked'. Place remains the individual space of a diffused (but explored) topology: converting space into place is one's only way to emplace oneself in one's topology – a sort of 'microtopia' as Goodrich reminds us, that imposes itself on the grid of New York City.[67] Place is always a slice of space, and it can always diffuse its boundaries and dissolve in space, regress to being what used to be if properly forgotten – except of course for the emplacing ability of dreams that *placate* nostalgia by offering forgotten places as present enjoyments. *'To placate'* is the process of converting space into a place. The double entendre can be explored: by rendering a boundless space a personal place, we delimit our anxiety in the face of the vastness, we fight against our *horror vacui* by filling it up with memorabilia of our quotidian trips. We delimit our safety within walls of habit and we leave out the uncharted forest.

Placelessness, therefore, is what is left when place withdraws from space. To put it differently, in placelessness one returns to a topology before placation, to the luminosity of space *before* geometry, *but only after* in a continuation of recursivity. This space may be the locus of the Deleuzian virtual, or indeed the Lacanian real: whatever it is, it includes, centrifugally, the movement away from it, the withdrawal of place. So, although the absence of place is restituted, the focus has to be on the *re*-turn, the change of direction, from the usual colonisation of space (that is, its placation, potentially for legitimate reasons or reasons of legitimation), to the deterritorialisation of space. The movement is one of withdrawal from a space of ignorance, leaving the space *tabula rasa* to accommodate the nostalgia of place. Space is changed, no doubt. But so it would be: the continuum/rupture between the inside/outside (here: place/space) is constant, and in its oscillation it annihilates whatever is built on either side. The movement is away from space, but the turn, the crossing through the stomata, is what constitutes placelessness: it is a space before ever having been placated, but only once placation has withdrawn. Space becomes what Agamben calls 'irreparable': 'these things are consigned without remedy to their being-thus, that they have precisely and only their *thus* . . . they are absolutely exposed, absolutely abandoned.'[68] Abandoned

---

67 Goodrich, 2006.
68 Agamben, 1993:39. Irreparable, as part of Agamben's messianic thinking, is the condition of things after the universal judgment. Thus, irreparability can also be the condition of things before the possibility of judgment, as the moment that institutes necessity and contingency – the two 'crosses of Western thought' that disappear after the universal judgment.

and exposed, space and its things surrender to the tide of place, resplendent
in their 'incorruptible fallenness' while 'above them floats something like a
profane halo'.[69] But what does one do with a halo? Jump in it or frisbee it
back? A halo is the beacon of 'the individuation of a beatitude, the becoming
singular of that which is perfect',[70] and as such it signals the singularity of the
form, the singular in the plural, whose plurality is absorbed in its inoperability. A halo is a hoop of transition (one that leads nowhere but itself, no
doubt), the luminous edge of the pore, the crossing itself: God does not
need a halo, humans cannot have one (yet). Singularity and plurality brought centrifugally together in a perfection of unworkability, inoperability,
placelessness, crossing.

Porosity dissolves the division between space and place and introduces in
its stead the crossing: 'The truth is that I am happy only between the place I
have just left and the place I am going to. I am happy only when I am
travelling; when I arrive, no matter where, I am suddenly the unhappiest
person imaginable.'[71] Thus I depart, in full conscience of both mine and my
destination's substitutability. This departure, the withdrawal, the placelessness, is to be the crux of the present community formation. Enabled and
hindered by porosity in equal measure, withdrawal understands community
as the home one leaves behind, the placelessness of the crossing, the forgotten
knowledge and the abandoned ignorance. A community conceptualised at a
moment of placelessness, a community of crossing, of sharing the withdrawal from absence, is the kind of community that plays along continuum/
rupture. Community is a boundary selection, that is, a selection of porosity,
of a process that enables placelessness rather than identity, withdrawal from
marking rather than marking.

This is the community that environmental law abhors, while desiring.
Idem for the city. A community of uncertainty, on the boundary, cheating
on both system and environment while residing on neither, accentuates the
pain of the return to the form continuum/rupture with which the systems
operate (despite their descriptions) in relation to their absent environments.
The paradox is reinstituted without promise of Ithaca – just return. But
this is all there is: the only understanding of community that environmental
law and the city can have is one that oscillates between the singular and
the plural, the conscious and the social, the thing and the observer, without
marking any side but the crossing itself. Anything else is an imposition –
even the obliteration of community considerations falls under this category.
Community exists to the extent that its absence is present, variably as
nostalgia, simplification, return, fear; but also as desire, withdrawal, loss,

69 Agamben, 1993:40.
70 Agamben, 1993:54.
71 Bernhard, 1988:58.

'unworking'.[72] As it stands, community for environmental law is an intellectualised nostalgia for 'before', a liberal belief in collectivity and inclusion, a marginalisation of whatever is non-human. But it can also be the place to think about the environment of the law, conceptualise its absence, connect with it in a way that initiates and relies on withdrawal, and describe its fear of/desire for continuum/rupture with its environment. Likewise, community for the city is presently a simplification of the urban *durée*, a signifier that excludes the city in its elusiveness. But community can be released from the nostalgia of the local, and seen as the place of facing the absence of the urban environment. It can be understood (and employed) as the level where the continuum/rupture with the trans-systemic is conceived. It can be 'seen' as the event of absence itself.

To this effect, a short foray into critical community thinking is undertaken here, as exemplified by Derrida, Nancy, Blanchot and Agamben. A description – and I must apologise now for the inevitable inadequacy of such a description – of some of the main issues connected to the *chagrins* of community will then be put together in an attempt to delineate a space for ecological community considerations that deals with the usual ecological communitarian problems while suggesting ways of connection with the conceptualisation of environmental absence, inevitably through an underlying but twisted Luhmannian perspective. As a frame through which to begin the excursus, I am suggesting the Derridean concept of *supplementarity*. Supplementarity is the paradoxical co-existence of both supplement and supplant, an addition and a replacement. What is supplemented (in both senses) is the absent centre of any totalising field: 'If totalisation no longer has any meaning, it is not because the infiniteness of a field cannot be covered by a finite glance or a finite discourse, but because . . . there is something missing from it: a centre which arrests and grounds the play of substitutions.'[73] In every field, every text, every community, there is an absence right in the centre, both a central absence and an absent centre, around which supplementarity plays, always pointing to the incomplete, the boundary, the exit, but also the centre, the desire for completeness, the origin of desire. The relation between addition and replacement is ambiguous and always undecidable: the supplement can be at the same time 'a plenitude enriching another plenitude, the fullest measure of presence', or a supplement that 'supplements . . . adds only to replace . . . represents and makes an image . . . its place is assigned in the structure by the mark of an emptiness'.[74] Supplementarity confounds the limits of presence and absence, as well as the classical opposition between finitude and infinitude, bringing the two

---

72 'Unworking' is Blanchot's term, explained as abandonment, excess, idleness; see also below, note 81.
73 Derrida, 1976a:289.
74 Derrida, 1976a:144.

together in a contingent double-bind that builds on an absence in the centre of presence.[75]

In one of its manifestations, supplementarity is Derrida's retort to Hegelian *Aufhebung*, the absolute community in the sense of the final synthesis where the Spirit prevails. As Drucilla Cornell points out, *Aufhebung* for Derrida is simply another way in which alterity is subsumed to totality, where dissent is eradicated and the communion to the Spirit is infinite.[76] In that sense, supplementarity 'is *nothing*, neither a presence nor an absence . . . it is the very dislocation of the proper in general'.[77] The 'proper' is the supposed centre, the essence of both individuality and collectivity, the space that is removed by its supplementation. For Derrida, there is no community that can resist the interruptive force of *différance*, the chain of supplementarity that exposes the absence of centre. In the midst of every community there is a question: 'A community of the question, therefore, within that fragile moment when the question is not yet determined enough for the hypocrisy of an answer to have already initiated itself beneath the mask of the question, and not yet determined enough for its voice to have been already and fraudulently articulated within the very syntax of the question . . . A community of the question about the possibility of the question.'[78] This is neither the ideal community, nor a dystopia; neither a possibility nor a project – this is community in all its 'reality', a signifier in whose heart lies the paradox of absence.

Supplementarity allows placelessness to be revisited in a different light as an addition to and replacement of the absence of space with the movement of 'leaving-behind' – a negative, empty addition that returns space to the form of place/space, where no centre is to be found except as a void. In his *Inoperable Community*, Jean-Luc Nancy talks specifically about community as the retreat from the 'fulfilled infinite identity of community', that is, the illusion of universal communion in a superior totality (as opposed to community)[79] that knows no boundaries and includes everyone, without reserving

---

75 Derrida, 1973:101. Derrida's discussion of supplementarity emerges from his reading of Rousseau. Perhaps the best way to understand supplementarity, destined to remain in a footnote, is masturbation, and especially Derrida's retort to Rousseau's guilt for much too frequent masturbatory sessions. Derrida, 1976a:154, argues that 'it has never been possible to desire the presence "in person", before this play of substitution and the symbolic experience of auto-affection'. Rousseau's evocation of the absent Thérèse is what supplements the absence with a very present absence, thus enabling actual presence in sexual relations. As an ultimate expression of supplementarity, Woody Allen in character in *Annie Hall* (1977) said 'Don't knock masturbation! It's sex with someone I love.'

76 Cornell, 1992b.

77 Derrida, 1976a:244.

78 Derrida, 1978:80. See also Gaon, 2005.

79 For Nancy, 1991:28, communion is sublation to a totality superior to their individuality and politically equated with fascism.

a place for difference.[80] To situate it in the present discourse, community is to be found in the withdrawal from a space of infinite knowledge and infinite connection, back to a place of systemic enclosure. Community is neither out there, nor in here, neither in the environment nor in the system, but fluctuating right in between: a mobile clustering around a contingent boundary between individual and collective, knowledge and ignorance, desire and fear, but with a specific direction away from space, away from infinity (but only after infinity has been visited, and only with the understanding implicit in the withdrawal that the direction will change again towards infinity in an oscillation of identity construction based on difference). This retreat from infinity, Nancy (after Blanchot) calls *unworking* ('désoeuvrement'), a notoriously difficult term to explain, which variously refers to abandonment, mourning, excess, idleness, and opposes itself to 'works', production, utility.[81] Unworking here can be understood as the retreat from 'works', from the exertion of colonisation, the fight for universal, the need for linkage either in the form of identity or difference. In the withdrawal from works itself, singularity is interrupted, community is interrupted, future is interrupted, and the *continuous interruption* provides for the paradoxical community of continuum/rupture. Retreating from space is an unworking of thematisation, of mastery and power claiming: in many respects, an autopoietic system's closure is the unworking itself, in that it presupposes a cognitive openness (movement towards space) that enables closure (movement away). Unworking is 'the blanks and the rips and tears of daily communication' that lead to the 'violent dissymmetry' between the self and other.[82] There is unworking at play 'in the positive labour of identification that binds any social institution to an insufficiency it cannot control'.[83] Unworking, here, is the affirmation of closure, a retreat to the absence of community, to the absent community inside and to a ground on which community is thought in terms of its continuum/ rupture with the environment – an environment that remains absent in its inclusion in the system.

If the environment can be simplified as the systemic desire for/fear of community, it becomes perhaps more obvious how the environment (in its absence, or rather in its tamed apparition) is included in the system: Nancy talks about the *clinamen*, namely an inclining towards the other, included in the self.[84] This clinamen is the bringing together (the *in* in 'in common') and the keeping apart by virtue of its being a space within each self, separate from

80 Nancy, 1991:xxxix.
81 'that which, before or beyond the work, withdraws from the work, and which . . . encounters interruption, fragmentation, suspension'. (Nancy, 1991:31); see also Joris's (translator's) preface to Blanchot, 1988, esp. xxiii-xxiv.
82 Madon, 1998:65.
83 Iyer, 2001:70.
84 Nancy, 1991:3.

alterity. It is my understanding that clinamen must remain foreign and proper, continuous and ruptured in its role as included exteriority: 'within its very separation, the absolutely separate encloses, if we can say this, more than what is simply separated . . . the separation itself must be enclosed, the closure must not only close around a territory . . . but also around the enclosure itself.'[85] In *Being Singular Plural*, Nancy pronounces that 'the outside is inside', but 'its negativity is not converted into positivity'[86] – it retains its inoperability, its absence, its question at the centre of the 'interiority without an interior'.[87] In every connection there is proximity and distance, in every internalisation there is exteriority within: 'at the heart of the connection, the interlacing of strands whose extremities remain separate even at the very center of the knot.' Collectivity and singularity, without mediation and without commonality, except for the common absence of commonality. Through the concept of clinamen, Nancy explores his community which is neither an atomistic-contractarian nor a socialist formulation, but an inessential, relational sharing of *finitude*, which Nancy defines as 'the infinite lack of infinite identity',[88] namely the exposure of the illusion of universal collectivity, the bringing forth of thresholds of difference, the retreat to the absence of identity – and this very absence of identity is what is shared.

The most consistently absence-relying formation of community comes from Maurice Blanchot's *The Unavowable Community*. In the heart of Blanchot's community there is negation: a negative community that defines itself in its negation by the other, by the one with whom the very community is expected to be formed. Blanchot returns to the question at the centre of the community, by relating it to the principle of incompleteness, the platonic ideal reinterpreted to seek, not fulfilment, but a solitary clustering around a negation: 'The refusal is absolute, categorical. It does not argue, nor does it voice its reasons . . . Men who refuse and who are tied by the force of refusal know that they are not yet together. The time of joint affirmation is precisely that of which they have been deprived. What they are left with is the irreducible refusal, the friendship of this certain, unshakable, rigorous No that keeps them unified and bound by solidarity.'[89] The negation in the heart of community is the negation of the community itself, the impossibility to experience community except through death. For Blanchot, as well as for Nancy who also follows Bataille in his dealings with community, death of a member is what forms a community: someone else's death in my presence is

85 Nancy, 1991:4. Also p.6, where Nancy situates singularity not on the level of atoms, but on that of the clinamen. Interestingly, see Serres's, 2001, use of the same term, and in a manner that supports the simultaneity of foreign and proper.
86 Nancy, 2000:13.
87 Nancy, 2000:63.
88 Nancy, 1991:xxxviii.
89 Blanchot, 1997:111.

what calls me into question more forcefully, interrupting my illusion of infinity, and replacing it with the threshold between 'the infiniteness of alterity' and the self's 'inexorable finitude'.[90] Blanchot's community is one of absence ('a community of absence always ready to transmute itself into the absence of community')[91] and of negation of the very existence or utility of community. ('What purpose does [community] serve? None.')[92] The Negative Community is solitude shared around the absence of community, on the cusp of the 'violent dissymmetry between myself and the other'.[93] On account of this dissymmetry, community is exposed to its internal absence: community 'exposes by exposing itself. It includes the exteriority of being that excludes it – an exteriority that thought does not master, even by giving it various names: death, the relation to the other, or speech . . .'[94] This included excluded exteriority is precisely the absence around which community clusters, the absence of the centre, the question in which the paradox resonates – in other words, what I have earlier tried to explain in terms of *unutterance*: neither speech, nor death or alterity are descriptive of the paradox of continuum/rupture between the system and its included undomesticated exteriority, its relation to its absent environment.

Blanchot alludes precisely to the form continuum/rupture when he talks about communication and fissure between the self and the other – 'two moments that can be analyzed as distinct, though they presuppose each other by destroying each other'.[95] Continuum/rupture reappear in the ambits of community and link it without linking to an included exteriority of exteriority, a negativity that can never be employed, an absence that can never be invisibilised: 'community proposes or imposes the knowledge (the experience, *Erfahrung*) of what cannot be known; that "beside-ourself" (the outside) which is abyss and ecstasy without ceasing to be a singular relationship.'[96] *Community is the ground on which absence is thought*, the ground on which solitude measures up with its supplement (its extension and substitutability). On the same ground, ignorance and exteriority are accepted without being converted to collectivity or retreating to individuality: the 'centre' of the community is the very decentring of the centre, the absence that remains absence and expresses itself as the question of continuum/rupture.

Finally, on a less negative but more fragmented ground, Giorgio Agamben's *The Coming Community* is a performatively 'crazy, slightly drunk'[97] text.

90 Blanchot, 1988:17.
91 Blanchot, 1988:2.
92 Blanchot, 1988:11.
93 Blanchot, 1988:22.
94 Blanchot, 1988:12.
95 Blanchot, 1988:22.
96 Blanchot, 1988:17.
97 Wall, 1999:121.

Agamben's 'putting-together' of a community is but an explosion of *pensées*, drawing the reader in a certain 'belonging' but without enabling the identification of its conditions. A community of readers, scattered in their singularity of a *whatever* community: *whatever* frees from the obligation of choosing between 'the ineffability of the individual and the intelligibility of the universal':[98] 'common and proper, genus and individual are only the two slopes dropping down from either side of the watershed of whatever.'[99] Agamben's community is also one without essence, neither atomistic nor holistic, without attribute or identity; at the same time, singularity is *whatever* singularity, its 'being-thus', in an ipseity determined only by its bordering to 'the totality of its possibilities', to a horizon of potentiality to which the singularity belongs 'but without this belonging's being able to be represented by a real condition: belonging, being-*such*, is here only the relation to an empty and indeterminate totality'.[100] This belonging, the unrepresentable, empty, pure belonging, which does not relate to either a condition (a category to which one belongs), or to the mere absence of conditions (in the manner of Blanchot's negative community), is what determines community. The belonging to belonging itself, that opens up to the horizon while exposing substitutable, *whatever* singularities, is clearly not a return to nostalgia, but to a community that has never been – a messianic community, always to come, that confounds contemporary politics with its lack of identity. A community of singularities that does not affirm its identity, a co-belonging that is not represented by any condition of belonging, and cannot exhaust itself in opposition to the law or political protest, a community that proves its potentia by its potentiality not only to be, but significantly, to *not*-be,[101] is a permanent threat to the law, an irrelevance to the sovereignty, an exclusion which can be killed off in the presence of the law's suspension.[102]

Nancy, Blanchot and Agamben's communities recast the discussion on community of any sort, and offer ways of reconceptualisation of the traditional issues of nostalgia, geography, locality, difference and collectivity – all issues that appear regularly in environmental communitarian discourse. This is the focus of the next section, where ways of advancing the concept of community, already hinted at during the above excursus, will be contextualised in the discussion towards a coming community whose connection with its environment will be adequate to accommodate the complexity of

---

98 Agamben, 1993:1.
99 Agamben, 1993:19.
100 Agamben, 1993:66. Whatever has been described by Wall, 1999:131, as 'the delicate interval' between the named thing and its being named, which constitutes the 'pure *thusness* of being'. It is in that sense that singularity disappears not in but *through* language – see Bartoloni, 2004.
101 In the sense of not potential; Agamben, 1993:34.
102 Agamben, 1993:85–6; also, Agamben, 1998.

continuum/rupture, while remaining operational for environmental legal and urban purposes. The proposed community formation, although absent, void and located in the movement of withdrawal, is emphatically not a manifesto for a neo-liberal, individualistic kind of rejection of community. It is precisely because of its absence that community can exist, and it is precisely because of its being necessary that community must remain absent.

III COMMUNITY WITH THE ENVIRONMENT

Locating community within its environment while maintaining the exteriority of the environment as already constructed within the autopoietic system, requires a differentiated understanding of the relational. In marked difference to Blanchot's solitary focus and Agamben's radical diffusion, Nancy begins with an understanding of the relational not entirely devoid of nostalgia.[103] While this relational is indeed placed on an absence – that of essence, of identity, of communion – its interruption is always predicated on 'an originary or ontological "sociality" ', whose limits are determined by *sharing*. Nancy admits that 'it is not obvious that the community of singularities is limited to "man" and excludes, for example, the "animal" '.[104] Nor, however, is it obvious that it is *not* limited to 'man' – the need to pronounce the exteriority, not just as singularity but as fragmented singularity, or indeed, unbounded and trans-singular singularity, is arguably more than just a matter of 'not obvious-ness'. In other words, the supplement, in the Derridean sense, of Nancy's 'originary sociality' has to be taken into account and fleshed out. Singularity, in its singular way, is the stuff of a continuum (not communion, similarity or even difference, but continuum as the unavowed paradox of the boundary). The rupture of sharing (this is the supplement, in the sense both of the sharing that ruptures, and the rupture of sharing) can never be left out safely, for then absence is reduced to workability, to production of absent presence. What is more, its co-priority with continuum cannot be forgotten, however ethically one is geared. Nancy's understanding of the Heideggerian *mitsein* ('being-with') is one of ingrained, prior collectivity which determines the singularity from within but always in relation. There is no (co-)prior, recursive fissure except on the level of singularity.[105] Nancy's absence is fundamentally positioned in a framework of appearing – even more, of *compearing*, of appearing in common,[106] but only with difficulty of disappearing, appearing out of common,

103 Also expressed as melancholia: see Kellogg, 2005.
104 Nancy, 1991:28.
105 In the sense of the sovereign; Nancy, 1991; see Motha, 2005.
106 Nancy, 1991:28.

or not appearing. It lacks, in twisted Luhmann parlance, a 'disintegration' value.[107]

Singularity, as the potential ground for such deviant appearances, always takes place on the level of *clinamen*,[108] thus reinscribing itself to an identity of difference (rather than a difference of difference). As already mentioned, clinamen can be understood (without moving away from Nancy's understanding) as simultaneously proper and non-proper, a 'space' from which to withdraw, rather than a place of convergence. Thus, fear of the clinamen is conceivable also in other contexts apart from totalitarianism.[109] This is more than just an ecological argument – it refers to a fissure, experienced as a paradox, whose trauma exceeds the individual-versus-collective and carves spaces and places of continuum/rupture within each of these and beyond these. The 'beyond' is nowhere to be found but merely in the oscillation between place and space: withdrawal from space (Nancy's 'retreat' from common identity) as inoperability of community can work (better: 'unwork') only if the space has *already* been seen as the locus of fear of/desire for the continuum/rupture between inside and outside, and only with the promise of return, not because of the clinamen, but for fear of it.

Locating community on movement does not annul localisation, but recasts it in the light of the variety of space. The typical association between place and 'proper' (Agamben's other side of 'common') is relativised by the systemic topology, which is constantly modified through the processes of production and consumption of autopoietic boundaries. In the urban system, the intensification of the social/conscious demands a systemic topology of movement, of ebb and flow that can no longer be equated to constancy, tradition or habit – these too, but only as one side of the topology. In locating community on the withdrawal of place from space, an understanding of the need to visibilise absence emerges. To put it in Nancy's terms, withdrawal is recognition of the boundary between finitude and the lack of infinity – boundary in the sense of Agamben's 'threshold': not a limit but a point of contact with an exteriority that remains empty.[110] Blanchot's negative community offers the conceptual tools to perform precisely such visibilisation, since for Blanchot community is formed around the very absence of community.[111] It is the other, with whom the community is expected to be formed

---

107 Luhmann, 2002b, following Gotthard Günther, talks about a 'designation value', namely the value that designates what is. Disintegration value would be the value that designates what is not by negating it.
108 Nancy, 1991:6.
109 Where Nancy, following Bataille's aesthetic attraction to fascism, equates communion with totalitarianism. See also Iyer, 2001.
110 Agamben, 1993:66.
111 The difference between Nancy and Agamben, on the one hand, and Blanchot on the other, is that the former cluster community around the absence of identity, whereas Blanchot follows Bataille more closely in envisaging a community for the ones without community.

and who is already included in community as absence, that questions, contests and negates the community, thus inflicting the fissure that makes community 'decompose constantly, violently and silently'.[112] What remains of the community is the question at the heart of the absence, the absent centre that decentralises itself and denies any localisation or permanence. The 'decomposure' of community is nothing else but the departure of place from space, the opposite of placation, the flight to anxiety, the movement away from centralisation and into the oscillation of withdrawal.

The community of two, the community between the one and the one's environment, the community of one and the others: there is merit in transferring these terms into the context of autopoiesis. The gestures are already in place: the absence of the environment as correlative of the absence of the very community with the environment. It may be odd talking about community in the context of systems – there are other tools, such as interpenetration, structural coupling, observation and so on, all equally debilitating of convergence, all struggling with the implausibility of communication. Nevertheless, such talking does not depart from the autopoietic struggle – on the contrary, it reinforces the problematic of the boundaries. It also allows a view of the system as part of a community with itself – a mirror community of continuum and rupture, invisible even to itself. It also underplays the importance of society – not by engaging with its opposite, but by annulling the usual ideas about community's succession by society, thereby making them at best indifferent to each other and at worst eradicating the need for the conceptualisation of either. For, indeed, there is no need to deal with either of these – absence is already adumbrated. With the introduction of the community of absence, the gesture of withdrawal is introduced as a way of further depopulating absence, thereby making it more present for the paradox of unutterance. Simultaneously, community discourse points to a slightly risky but potentially enlightening use of the tools of observation and blind spot, allowing thus for a construction of collectivity that appends itself to the discussion on the prior invitation between environmental law and the city.

A community with the environment on the basis of absence is, first of all, a community that internalises its 'central' absence: if nature is no longer the host, if the human is no longer whole, if the object of environmental law is nowhere to be found but in the absence of its environment, if the city incorporates the ghost of its environment in its boundary topology, then community returns to a traumatised singularity, a fissured autopoiesis, a porous separation of difference and identity. Community is revealed as, to rephrase Nancy, the infinite lack of infinite communication, the appearance of unutterance in its blinding visibility. The system returns from its excursus into alterity empty-handed – and this is where community is to be sought: on

112 Blanchot, 1988:6.

the withdrawal from space and into the obverse of placation. Placelessness is the withdrawal from ignorance *but only after*. Not unlike the projection of risk into the future and back, placelessness denotes the loss after the loss, the flight from community itself and from the ideology of knowledge, and back (but not yet, and only until the next withdrawal) towards the knowledge of ignorance. An ecological community is a community that withdraws from its possibility, allowing absence to restitute itself in that space, to 'present' itself. Community is what I leave behind when I feel protected. It operates as the receding basis of my excursions, and at the same time the locus of my excursions. Veritably on the boundary, an ecological community in community with the environment turns its back to the environment and proceeds to placelessness, while the environment returns to its proper unutterance (always as an absence within the community).

This is the loss at the centre of the ecological community. Blanchot talks about the death that constructs the community while internalising its *arkhé* and its conditions of creation: no overarching principle, no authority or power pulls the community together, except for its own inability to be, its own loss.[113] The community with the environment can only operate along autopoietic lines of diffusion of power, of enclosure of responsibility within the very movement of contingency,[114] and of situatedness of responsibility within the broader spectrum of the paradox continuum/rupture. If the outside has proceeded to the era of post-ecologism, it is only fitting that the community returns to placelessness withdrawing from the space of the tragedy of the commons and making its absence felt in the village green.[115] Through the withdrawal from space, the community acknowledges the loss at its centre, the loss of the centre and *arkhé*: no longer Nature to embrace it, no longer legitimacy to pacify it, no longer illusions of continuum that bring everything together, or clear-cut rupture that tells us who is with us and who against us. Life without *différance* is well behind, an illusion that at best can operate only retroactively, and only as nostalgia for the community that has never been.

But this nostalgia is a powerful pull. The horizons of its temporality are, predictably, reversed, and nostalgia returns to its return and dreams of the future. Community is expected: 'community without community is *to come*, in the sense that it is always *coming* endlessly, at the heart of every collectivity', says Nancy.[116] It is in this manner that the concept of the coming community can potentially operate as theoretical basis for environmental law's conceptualisation of intergenerational equity. The concept is an ingredient of

---

113 Blanchot, 1988:13–14. This is the community of the *Acéphale* that demanded a sacrifice in order to begin (and dissolve in its absence).
114 Rasch, 2000; Haraway, 2004; Wolfe, 1998.
115 Blühdorn, 2000; Hardin, 1992.
116 Nancy, 1991:71.

the recipe for sustainable development as popularised by the Brundtland report, which requires that the needs of future generations will not be compromised while present generations' needs are met.[117] There is weak and strong intergenerational equity, the former commanding that the present generation passes on to the next generation the same total stock of overall resources, whereas the latter, not being satisfied with rounding up tricks, requires the exact resources to be passed on. Each approach has its peculiar problems.[118] Thus, weak equity poses the question of commensurability of resources as varied as knowledge, coal, technological advancements, aesthetics, and so on; strong equity, on the other hand, operates more as an ethical call than a practical demand in view of the non-renewability of some of the resources.[119] These problems enter the greater problematic of sustainable development, whose merits and exigencies deserve (or not), and have at times been assigned, a whole different set of analytical tools.[120] Sustainable development aside, the issue of future generations requires from environmental law an internalisation of a temporality that excludes the messianic. To put it differently, future generations are seen as part of present generations in terms of environmental law's operations, in that future generations are not just to come but already present.

Luhmann's construction of time, as I have shown in the previous chapter, is present-based, in that the simultaneity of operations between the system and the environment requires that everything that happens, happens in the present. While each system has its own temporality, as I have argued earlier, across systems there can only be simultaneity that manifests itself as *expectation*. In that sense, for Luhmann the future can never begin, and consequently it is more correctly referred to as 'future-present'.[121] While recursivity replaces origin, messianism is a thing of the past: 'time can no longer be depicted as approaching a turning point where it veers back into the past or where the order of this world (or time itself) is apocalyptically transformed.'[122] Any turning point is replaced by a present temporal extension of the system, thus reinforcing the simultaneity of operations. This formation is not fundamentally dissimilar to the Derridean concept of the future. While impressionistically the difference is unbridgeable, since for Derrida the future has already begun, on closer inspection one sees that neither of them expects anything from the future, since the future either

117 World Commission on Environment and Development, 1987; UNCED, 1992; Princ. 1 & 8, 1972 United Nations Conference on the Human Environment ('the Stockholm Declaration') UN Doc, A/CONF/48/14/REV 1, etc.
118 Beckerman and Pasek, 2001.
119 Alder and Wilkinson, 1999; Kavka and Warren, 1983; Weiss, 1989.
120 One of the most succinct and thoughtful analyses is by Stallworthy, 2002, see also, Redclift, 1991; Richardson and Wood, 2006; and Philippopoulos-Mihalopoulos, 2004b.
121 Luhmann, 1995a.
122 Luhmann, 1982a:272.

can never begin (thus expectations are contained in the present), or it is already here, so any apocalypse is reserved for and within the present moment.[123]

Derrida relates a story by Blanchot, in which a beggar recognises the Messiah at the gates of the city, and in his embarrassment as to what to say, he asks the Messiah, 'When will you come?'[124] For Derrida (as for Luhmann), the future can never begin, it is always yet-to-come. Of course, the future cannot be contained within the horizon – 'the opening and the limit that defines an infinite progress or period of waiting'[125] – but beyond it, located in an unknown futurity that cannot be circumscribed by expectation. Thus, the justice due to future generations, justice in its spectral totality, is messianic: justice 'does not wait'. At the same time, justice always remains elusive, deferred, the ghost of the undecidable: the yet-to-come of justice 'is not a horizon but the disruption or opening up of the horizon'.[126] Luhmann and Derrida meet at the gates of the city: while neither of them waits for the future, the future is just extramuros. The difference is that, for Derrida, a horizon is delimiting, while for Luhmann, who follows Husserl more closely in this, horizon is all that the system can become through its cognitive openness: the totality of systemic potentiality. But Derrida's subjects are thought to be cleverer than those of Luhmann's: the former formulate the horizon in terms of signification; the latter have no idea about what lies in the horizon unless converted into meaningful systemic communication.

Future generations are always already in the present, and there is no point in waiting for them. Except if one can wait without waiting, that is in the knowledge that Godot never arrives: 'Awaiting without horizon of the wait, awaiting what one does not expect yet or any longer . . . just opening which renounces any right to property, any right in general, messianic opening to what is coming, that is, to the event that cannot be awaited as such, or recognized in advance, therefore, to the event as the foreigner itself, to her or to him for whom one must leave an empty place, always . . .'[127] Waiting without waiting, fractally diffusing the presence to-come in every moment of the present, in every operation, in every simultaneity of presence: this waiting wraps little apocalypses in the folds of the system, sieves the apocalypse of the future through the porosity of the boundary, and fractally partitions it in shadows of angelic guardianship on the side of every legal gesture.[128] The law

123 *Contra* Cornell, 1992a, who, despite a careful play with Luhmann and Derrida, fails to see the multiple operation of the Luhmannian present, and misconstructs the systemic boundary as closure rather than closure/openness, thus equating systems theory with a rather flat presentocracy.
124 Derrida, 1981.
125 Derrida, 1992a:26.
126 Caputo, 1997:118.
127 Derrida, 1994:65.
128 Stone, 1996, for guardianship; Weiss, 1989, on ombudsmen.

sleeps with justice – no doubt contingently,[129] undecidably, spectrally, without ever touching each other, but certainly co-existing in some parallel presence/ absence, where future equity is presentified in its absence as the one who will never arrive. Present and future generations are placed together once again in a community of continuum, which ruptures itself around the absence of knowledge about the thing with which there is a continuum. And since everything takes place in the present, this community is messianic, waiting and holding back, while at the same time, moving, withdrawing – neither space, nor place, not even in-between, but the movement of place away from space, in a relentless recursivity that includes (while excluding) all futures in their blinding absence.

The community of continuum/rupture is not just intra- or inter-generational, but transgenerational. It trails the differentiated kinds and apparitions of generations, cuts across them by encircling absence instead of communion, thereby converging while interrupting. This *trans* across anthropocentricity and ecocentricity arrives (in recursivity) at a fractal point of central diffusion, where the -centric discourse becomes obsolete. In the *trans-*, with its interrupted immanent beyond, the ecological community describes itself (if it ever could) as the perpetual environment, the other side of space, and thus neither human, nor non-human,[130] neither present nor future, neither local nor global, neither immediate nor distanced.[131] The limits of this total absence are to be located in the presence of absence itself as internalised within the system. The system, be this environmental law or the city, internalises the absence of its community with the environment in an operational way that checks for cognitive openness, whose thematic, however, is in turn limited by the limits of the internalised absence. Systemic cognition is necessarily invested in and limited to the boundaries of the absence, in the form of a cognitive continuum and a normative rupture. Or indeed, a cognitive rupture and a normative continuum. However it is, the internal boundary of ignorance, the halo around the space of unutterance, breaks and mends simultaneously, withdrawing the community from any illusion of identity and awarding it with a *whatever* inoperability and a yet-to-come presence.

IV RIGHTS IN THE ENVIRONMENT

Rights are linked with the legal representation of community in various ways. Here, the focus will be on their commonality of location: rights and community are to be found on the withdrawal from space, both riding their

129 Justice is a contingency formula for Luhmann, 2004 – see above, Chapter 3.
130 Fox, 1990, on the transpersonal ecology.
131 Golding, 1972, on the need to prioritise the immediate future generation.

absence and instituting themselves as paradox. Thus, rights populate the space of absence in a paradoxical way that links without deparadoxifying the community and its environment. In their fundamental paradox of particular/universal, with its utterable shoots extending towards individual/collective and substantive/procedural, rights are adequately equipped to express the paradox of community as neither solitary nor collective, neither leading somewhere nor being static, but as the movement of withdrawal. Still, this section does not hope to be yet another version of the paradox of rights, but an expansion of sorts that enriches the paradox with a reference to the possibility of environmental rights, that is, variably rights of and to the environment, in their collective or individual articulation.

Two aspects of the issue of rights and environmental protection are relevant to the present discussion: first, the link between the exercise of environmental rights and the absent community. This goes beyond the discussion on whether environmental rights should be individual or collective, and attempts to contextualise the concept of withdrawal. Second, the debate in environmental legal literature on whether environmental rights should be substantial or procedural. Of course, the two issues cannot even schematically be separated since they pose the same problem from a marginally different perspective: the first asks 'Who is to exercise these rights?', whereas the second 'What rights are to be exercised?' Both issues stem from the same long-standing and well-documented debate that has divided environmentalists in opposing camps.[132] The central point of the debate is whether environmental law can in fact benefit from the protection afforded by the mechanism of legal rights or whether the looseness of the rights approach will eventually debilitate it. The most popular, and radical, way of linking the two is by introducing a separate right, solely dedicated to the protection of the environment.[133] The alternative suggestion is the radicalisation of existing sets of rights in such a way as to include concrete and guaranteed references to environmental protection. This does not necessarily involve new rights, although it certainly involves a specialisation of already established ones.

The above presupposes an answer to the question with regard to who exercises these rights. In other words, is it possible to separate rights from *human* rights?[134] As it stands, whatever rights are recognised to non-human entities, they will be superimposed on the interests of the latter as a cloak of good-will divination.[135] Stone foresaw the problem when he suggested that, in principle, anything can be a subject of legal rights, as long as there is

---

132 Boyle and Anderson, 1996; Shelton, 1991, 1992; Eckersley, 1996; Thorme, 1991; Sands, 1994.
133 For doubts, see Shelton, 1991, 2000; Kiss and Shelton, 1993; Hodkova, 1991; Sands, 1994; Shelton, 2003; but cf. Desgagné, 1995.
134 Elder, 1984.
135 E.g., Sagoff, 1974, 1988; McClymonds, 1992.

somebody to represent it.[136] This suggestion is a developed form of his initial thesis, namely that trees, valleys and other natural entities should be conferred rights.[137] The debate relies on the division between human and non-human and reiterates the anthropocentric/ecocentric arguments. The problem with this division is the assumption of rupture, which can be mended through the administration of continuum. Thus, rights as the legal expression of the continuum. Considering the improbability of the suggestion, one wonders why it is so conservative. The exclusion of rupture from the 'normal' course of things is intensely problematic because it fails to capture the complexity of the human/non-human connection. Even if one were to accept that rights are an adequate way of addressing the connection (and in this post-Århus era, this is disputed only with great difficulty),[138] rights ought to be explored in their full paradoxicality before any margin for such connection is presumed.

The issue is intermeshed with the debate on individual and collective exercise of environmental rights, since more often than not the litigant who finds it difficult to prove interest is an environmental group. Theory is divided between, on one hand, the need to accommodate the practical preference of the legal system to the individual over the collective,[139] which has led several authors to reject the validity of contribution of the rights mechanism to the environmental cause altogether;[140] and on the other, a considerable tendency to favour groups as the most appropriate means of integration of the socially[141] and ecologically marginalised, especially by appealing for the general acceptability of *locus standi* of non-governmental organisations, and the broadening of the criteria for standing, particularly the amplification of

136 Stone, 1996.
137 Stone, 1974. This approach was fraught with difficulties, such as the issue of equality of rights, which can lead to rights of viruses. Since then Stone, 1987, 1996, has advanced his theory, by suggesting human guardians. This in turn raises the problem of *locus standi*.
138 The Århus Convention [Convention On Access To Information, Public Participation In Decision-Making And Access To Justice In Environmental Matters, Århus, Denmark, 28.6.1998, 38 *International Legal Materials* (ILM) (1999) 515] has linked environmental and human rights, although without proceeding to rights *of* the environment specifically or substantive environmental rights generally; see Brady, 1998 and http://www.unhchr.ch/ environment/index.html, accessed on 1.3.06. Also, Shelton, 1992, on how environmental rights do not fare well on the human rights long list. For a more positive perspective, Boyle, 1996. It is, however, worth mentioning that dispute about the connection has come from an unfortunate source: the Johannesburg Declaration, [Report on the World Summit on Sustainable Development (Johannesburg, South Africa, 26 August–4 September 2002) UN Doc. A/CONF.199/20), Resolution 1, Annex, at 1–5], which only with great hesitation endorsed that it is *possible* for environmental protection and human rights to be connected. See Pallemaerts, 2003.
139 Most domestic legislation reserves these rights for individuals: see S. Grosz, 1995.
140 See Alder and Wilkinson, 1999; Merrills, 1996.
141 Lacey, 1998.

terms such as 'victim', 'sufficient interest', etc.[142] Simultaneously in the affirmative and the interrogative, the environmental movement describes itself as 'concerned to "represent" the interests of a considerably expanded constituency of "non-citizens" who cannot vote or otherwise participate in the formal political decision-making processes or territorially defined polities, but nonetheless can be profoundly affected by the decisions made in any given polity'.[143]

The obvious problem with the existing legal approach is the narrow definition of the interested party. In a legal system, where access to the judiciary is organised in terms of individual/collective, the margins for environmental representation are significantly reduced, for quite often the represented cannot appear as representative of anything from within the legal system – unutterance represents nothing and is represented by nothing. Practice shows that individual representation is bound to private property rights, that only with great difficulty can they cater for the collectiveness of general environmental concerns; on the other hand, collective representation often lacks the linking element of personal interest.[144] In whichever way the issue is approached, the concept of environmental protection is subjugated either to the protection of personal interest, which only in an ancillary way includes the environment, or to antagonistic endopolitics of non-governmental organisations that vie with each other as to who may be more appropriate to defend the particular environmental cause.[145] The problem with both formulations is that the question of priority between ecological and other issues is not solved where it ought to be – that is, legally speaking, before the court – but much

---

142 See Faulks and Rose, 1996; S. Grosz, 1995; Fordham and Cameron, 1994; Desgagné, 1995; Harding, 1996; Bowden and Lawrence, 1994. In England, see *R v Secretary of State for the Environment, ex parte Rose Theatre Trust Co* [1990] 1 QB 504, where it was held that the Trust did not have sufficient interest to challenge a decision of the Secretary of the State for the Environment not to consider the remnants of the homonymous theatre as an ancient monument. It was held that 'a greater right or expectation than any other citizen was required', although it was also acknowledged that some decisions that affected everyone and no one in particular may be immune from challenge by judicial review; see also *R v Poole Borough Council ex parte Beebee*, and *R v Swale BC ex parte Royal Society for the Protection of Birds* [1991] 1 PLR 6.

143 Eckersley, 1996:354.

144 See Hardin's, 1992 (original 1968), famous *The Tragedy of the Commons*; also Stallworthy, 2002. It is, however, of relevance that, at least in England, the judiciary is progressively moving towards blurring the distinction between individual and collective for environmental protection purposes, focusing instead on proving a geographical link between the litigant and the locality of the perceived environmental threat, which automatically operates as a presumption of interest. See *R v Pollution Inspectorate ex parte Greenpeace (No 2)* [1994] 4 All ER 329 where an environmental group, and *R v Somerset County Council ex parte Dixon* [1998] JEL 161 where an activist was granted standing on the basis of their geographical link with the Sellafield nuclear processing plant in the first case, and a quarry in the second.

145 Alder and Wilkinson, 1999.

earlier, at the stage of the *locus standi* decision, at which stage substantial environmental arguments are condensed to arguments of linkage.

So where to locate the *locus standi* of an absent community? What is the disruption of a *locus standi* in absentia? Where does the law turn when its re-turn to its centre is a turn to its absence? Environmental law internalises the absent community as a continuous interruption of its operations. In its community with the environment (in the form of *trans-*), an absent community is continuously present in the system as an absence around which the community clusters. Can an absent community have rights? If it can, they will be well beyond the human/environmental rights discourse, since the distinction does not hold in an absent community. The rights that can be employed by an absent community are located on the same movement of withdrawal, and in the paradoxical locus of always-the-other-side of the movement. Only thus is the space needed for the presence of unutterance revealed. The example of an urban community that perpetually undermines itself in its boundaries and membership, without anchoring on any of the extremes that the law comprehends as indicative of a right-holder, brings forth the spaces of unutterance within the law and in its coupling with the city. In urban spaces whose relation of continuum/rupture with their environment (of present or of future) is visible, communities of absence are conceptually possible, and rights in the environment – namely, rights always on the other side – makes such conceptualisation imaginable.

Rights have always-already been conceptualised within the operations of the system. They have been 'recognised'[146] (in the sense of something that has been, latently or visibly, present all along) at an arbitrary moment which subsequently has been arbitrarily hailed as the point of modern differentiation.[147] Before differentiation, the premodern: the hierarchical organisation of compartmentalised subjective status, whose overcoming was practically impossible. One remained in the social position in which one was born, melting away in the hierarchical delimitation of one's origin and, hence, one's future. 'Subjective' rights, in the sense of 'of, and because of, the subject', were superfluous in such immovable social conditions since individuality and social position were identical.[148] This was a position of total inclusion, where the individual was absorbed by hierarchy. After differentiation, the modern: subjective rights emerge as a vehicle that enables the 'subject' to pick and mix social positions, thus reinstating the universal impact of human rights while emphasising the possibility of subjective differentiation. Human rights are the compensation for the loss of a relatively stress-free yet claustrophobically fixed social position: they guarantee the *exclusion* of the human being from

---

146 Also in the Hegelian sense: Douzinas, 2002.
147 Luhmann, 1995a.
148 Luhmann, 1999; Verschraegen, 2002.

societal communication, thereby allowing her the opportunity to choose.[149] This exclusion is seen as the much-needed rupture of a regime whose continuum allowed no space for margins.[150] At the same time, human rights are the mechanism that maintains modern differentiation by preventing any one system from colonising another. Through the maintainance of exclusion (through what I have called 'rupture'), the continuum of differentiation is enabled to carry on. This paradoxical description of human rights as the 'grid' that both separates systems and keeps them this way, and the 'vehicle' that circulates on the grid and allows for the individual to alight at any of the said systems without being fixed to any of them (in other words, the rupture that enables the continuum, and vice versa), recasts the unutterable paradox of universal/particular in a new form that alienates rights from their individual location and looks at them in their social dimension. One only has to take the above formulation and throw it into its environment: thus, rights are the tool with which the individual withdraws from inclusion, and systems become aware of the absence of the other side of their differentiation: their blind spots. In both cases, rights reveal the absence of the other side, of an environment increasingly closer to becoming proper, but at the end always slipping away.

It is, then, not a great disappointment, when, in view of the above, the division between substantive and procedural environmental rights is deemed superfluous.[151] At first instance, substantive rights cannot be disengaged from continuum. They are collective par excellence – not in the sense of enforcement, but in the sense of recognition: it is only in community that a solitude dons a substantive right, its part in the (even ecological) *humanum* thereby recognised, for otherwise the grid would collapse in an all-thematising dystopia. Admittedly, the collectivity of rights is a ruptured collectivity of ruptured solitudes. But the collectivity of rights is the only form in which rights can be described. Rights of solitude cannot be described, and solitude cannot hold rights – except if in community. But the undescribability of the right, the absent side of the right that is presupposed for the presence of right, remains of the solitude, it cannot be of the community, it can never be shared: if it is, community is converted into communion, a 'work', a production of itself as the antithesis of solitude. The paradox ought not to be resolved. Rights are neither of solitude nor of community, neither of continuum nor of a rupture, neither of exclusion nor of inclusion. Only in absent community (the community that withdraws) can solitude balance on the *threshold* of rights,[152]

149 Luhmann, 1997a.

150 Philippopoulos-Mihalopoulos, 2006. See King and Thornhill, 2003, for an excellent discussion on the liberal subject in Luhmann.

151 For the debate, see Boyle, Douglas-Scott, and Cameron and Mackenzie, all in Boyle and Anderson, 1996; also Shelton, 2003; Krämer, 2000; Kiss, 2003.

152 Agamben, 1993:66.

which is nothing else but their unutterable paradox: neither objective nor subjective,[153] neither particular nor universal, neither of the self nor of the other,[154] neither boundless *jouissance* nor bound oppression, neither a utopian telos nor a lie,[155] neither immanent nor transcendent.[156] But *nor* are they *neither* of these: rights follow the withdrawal from space, the movement of the community from a cognitive totality of identity and *ennui*, to the flight away from itself, from community as a basis of 'work'. Rights *are* one of these 'neither-nor', but *only after* and only in withdrawal. The space of rights can only emerge in its full paradox, when place is holding back.

The bringing forth of solitude is no more of a rupture than Nancy's singularity.[157] It is indeed a rupture of sorts, but one contained in the continuum. Rupture can only be inflicted in parallel, simultaneously and in a form with continuum. If not, community proceeds to the trenches of individual/collective and loses its absence. Community is ruptured to experience the loss that makes the community. This space remains empty of solitude (recognition only in community), yet full of the absence of community (the right is of the solitude but after the withdrawal from space, and then *only after*). The locus of rights, therefore, is empty. No-one can stand there without risking total inclusion. Rights are perpetually exiled in the environment of certainty, in the space from which withdrawal has just been performed. The space of recognised rights can only be populated by the departure of community – the place of singularity withdraws from space and marks the threshold between outside and inside. And there it would seem – again impressionistically – that the movement that brings the rupture of the continuum is the procedural right. Away from universalism/relativism, the rupture is not equated with the marginalised 'other' of a universal belief,[158] a supposed value-free proceduralism,[159] a smooth transition to environmental efficiency,[160] or an imagined opportunity for a level playing field; but the vehicle that takes the right-holder away from the self (as recognised in community) and onto the crossing. A procedural right links without subjectifying, operating on the basis of absence of subject and community, thus

153 Luhmann, 2004:415.
154 Thus Lévinas, 1997:176, '[Human rights express] . . . the alterity of that which is unique and incomparable, attributable to the suspension of every person in humankind, which ipso facto and paradoxically, abolishes itself so as to leave each human being unique in their own genre.' (my translation).
155 In his psychoanalytic interpretation of rights, Douzinas, 2000a, following Cornell, 1995, herself using Lacan, calls the locus of rights the 'imaginary domain'.
156 Motha and Zartaloudis, 2003, on Douzinas, 2000a.
157 Zartaloudis, 2002.
158 Derrida, 2005; Eckersley, 1996, for pro and contra; also Birnie, 2002; Boyle and Anderson, 1996, and Lau, 1996.
159 Habermas, 1996.
160 As famously argued by Ksentini, 1994.

eating into the *trans-* of the community from the inside, 'unworking' the community by exposing the absence of subject in the community, and the absence of community in the community itself. The procedural 'access to . . .' (justice, participation, information) stops exactly on the access: an excessive access, an access 'which, as workless, as pure excess, achieves nothing; but this is its very force'.[161] Its force lies in the absence of categorising, distinguishing, defining thresholds, subjectifying,[162] but also in the absence of grouping, uniting, finding similarity and recognising communion. Its force also lies in that it ends nowhere but to another movement, perhaps back to space, perhaps deeper away from it, but always in oscillation and always in recursivity with wherever substantive rights think they are going. This is the rupture that guarantees the continuum (and in that way, its own continuity). In any case, the movement can only be determined by the space it leaves behind. Following the movement of rights in their paradox as both procedural and substantive, one is led back to (the absence of) community. If an absent community is the only community possible in the context of environmental considerations, then a paradoxical understanding of rights in their movement between their incompatible extremes is a candidate way in which this community can be brought forth and understood by the law as from within the legal system. To put it plainly, environmental rights, in their diffused mobility, enable the much needed return to community. Environmental rights can now be seen as the movement, in its full circularity, between the individual and the community. Through rights, the community becomes unworked and its absence visibilised – and for this, necessary.

The unworking of an absent community nests in the blind spot of the observer. To recall the discussion in Chapter 3, every observation is a distinction. Second-order observation, the observation of *how* rather than *what*, visibilises the paradox of distinction of first-order observation. The unmarked side of a distinction includes the observer: 'one thing the observer must avoid is wanting to see himself and the world. He must be able to respect intransparency.'[163] The observer cannot see *the* unity – she can only see what remains after the unity has been severed. What she sees may well be *a* unity, but it will not include herself. She remains in a *blind spot*, namely the point of observation that enables observation to take place. The unity of the first-order observation (the unity that includes the observer and the observed) can only be observed by means of a further distinction, that is, via a different observer, who will also, however, operate from his blind spot. First- and second-order observation necessarily function together: 'observation is

---

161 Hannafin, 2004:7.
162 Boyle and Anderson, 1996:9, say that procedural rights provide a 'more flexible, honest, and context-sensitive approach'. Yes, if one understands by this, an 'undefinable, unmediated and decontextualised approach'.
163 Luhmann, 1998a:111.

possible only in a recursive network of the observation of observations, not in the form of a singular spontaneous, "subjective" act.'[164] Such a recursive network is formed on the distance between a second-order observer and another observing the former observing, and so on *ad infinitum*, or more accurately in a circle which involves ebbing second-order observers who fluctuate between immersion in and distance from (the observed, observing, self-observation, other second-order observers), and first- and second-order observation.[165]

Observation as a right? Neither just of the observer, nor shared amongst the network, observation is to be located in neither of these, but, unsurprisingly, in its very blind spot, its withdrawal from both solitude and community. Just as the blind spot is the ground on which the invitation between environmental law and the city takes place, in the same way it is the ground on which the absent community materialises itself. The blind spot is an interiority of the system, its shadow as it were, inaccessible to the system itself. At the risk of a critical clash, one is reminded here of Nancy's *clinamen* of singularity, the inclining towards the other, included in the self.[166] Clinamen is a foreign interiority, a gravitational space within singularity, just like a blind spot. In its clinamen-like nature, the blind spot pulls the observer towards the other observer, or the other observer towards the first observer, but without ever dissolving its singular limits and becoming one with the other. In that sense, the blind spot is the invitation – an absent, empty, inaccessible invitation. The blind spot is the autopoietic locus of rights, accessible by neither solitude nor community, but only in the movement away from either and towards the other. Luhmann admits: 'this imaginary space [i.e., the blind spot] replaces the classical a priori of transcendental philosophy.'[167] And further: 'the systemic keystone of epistemology – taking the place of its a priori foundation.'[168] And somewhere else: 'the blind spot is [the observer's] a priori, *as it were*.'[169] The blind spot replacing the a priori, the origin, the foundational phrase, the first gesture, the paradox of unutterability; but only '*as it were*' – not really. It cannot be *it*, because *it* does not appear in the system. The blind spot remains invisible, for otherwise one risks the rupture; not the rupture of monadology, of systemic closure and normative recursivity, but the rupture of that which comes from the visibilisation of absence, of unutterability.

164 Luhmann, 1998a:111.
165 The circle or network of observation has been suggested by Luhmann, 1993b (2004), as an alternative to Teubner's, 1993, hypercycle. See also my concept of *unbracketing* in the context of Husserlian intersubjectivity and community formation, 2001.
166 Nancy, 1991:3.
167 Luhmann, 1994b:21.
168 Luhmann, 2002d:136.
169 Luhmann, 1994a:28, my emphasis.

So, even in Luhmann there is a moment when absence is presentified. This is the moment of community (*as it were*, since the word is never used), when the phenomenological *here* becomes *there*. At this point, Luhmann's connection to Husserl becomes relevant. Husserl's community of monads,[170] especially in the way it appears in the Fifth Meditation, constructs the other's subjectivity as part of the self's peculiar ownness: the other ego appears as an object of my consciousness that shows itself as being *there*. However, the other ego is presented to itself as belonging to its absolute *here*.[171] The passage from my 'there' to the other's 'here', and the bringing together of the two as one alter, is possible on the basis of intentionality which does not posit two distinct spheres but one intentional community with other beings. This community is based on the possibility of the ego's experience of something that, although constituted in the ego, is nevertheless experienced as 'other than the ego'.[172] In consequence, every other monad constitutes the same community as the ego, but in its own intentionality and within its own monadological limits. Thus Husserl: 'within the vitally flowing intentionality in which the life of an ego-subject consists, every other ego is already intentionally implied in advance.'[173]

In this schema of interpenetrative egos, where the other (the environment, the coupled system) is apprehended internally, rupture is inserted in the form of reversal: on the basis of the blind spot, autopoiesis can now talk about the coupling of environments, the interpenetration of horizons, and the community of absence. The other is experienced as 'other than the ego', but through a 'there' that is never 'here' – a 'there' that remains inaccessible to the self, always hidden, always elsewhere than the self, yet correlative to the self's 'here' as the alter's 'there': this is the blind spot of intentionality, and the beginning of the absent community. The community of the blind spot is the community of second-order observers, a solitude that withdraws from community, turning its back to its being observed and thus abandoning itself to a construction of collective absence. This is where Luhmann's breath becomes visible: his proposed solution to paradoxical paralysis is a circuit of second-order observers.[174] But this is precisely where a community of continuum/rupture, of 'turned backs', and of withdrawal from observation is formed. There is no organisation, no communion of monads, and no communication about it; just the absence of observing systems, the withdrawal (wilful or not, makes no difference) from being observed but always

---

170 Famously criticised by Derrida, 1989, on the basis of priority of writing, and in terms of intentionality and monadology.
171 Husserl, 1973a.
172 Husserl, 1973a:128.
173 Husserl, 1970b:255; this is what Kockelmans, 1994, calls 'transcendental intersubjectivity', and Steeves, 1996, 'transcendent community'.
174 Luhmann, 1990d.

visible in that withdrawal, always exposed by the production of its own blind spot in its oscillation between the 'here' and the 'there', the first- and second-order observation. It is there where the 'silent foundation' of rights can be located[175] – in the very silence before the unutterance of the 'there'. This is a community of unworking: an absent, negative community that celebrates its confused movement, its perpetual oscillation and its construction through its very absence, 'the absence of the work as it *produces itself* through and throughout the work'.[176] The unworking shadows the systemic operation and poses little fractal eruptions of messianicity at every point in the temporality of the system, an interruption that is continuous and therefore invisible, and that 'neither destroys nor recreates but leaves the present not quite what it was'.[177]

This is reminiscent of Agamben's messianism:[178] in his analysis of Kafka's parable 'Before the Law', where a countryman is forever and repeatedly barred by the doorkeeper from going through the gate into the Law (a gate reserved exclusively for him),[179] Agamben reads in the seeming inaction of the countryman the overturning of the law which, although in force, is without significance: 'how something really has happened in seeming not to happen.'[180] The Law's redemption comes precisely in this act of overturning,[181] thus not requiring a replacement with a new law,[182] but contenting itself with the final arrival of/to what Agamben calls 'the historical and wholly actual homeland of humanity'.[183] A messiah that has arrived from the back door, already here in the other's 'there', and for this, inaccessible, invisible, absent. A home already here, reserved in blind spots of unsuspected observers: a home always to-come, and for this, the home that one needs to leave behind. This is the community that can embody rights that are also *always-to-come*: rights as movements of a finite community that retreats from the illusion of an all-observing position and is already in search

175 Perrin, 2004.
176 Maurice Blanchot, 'The Absence of the Book', 1969, original emphasis, cited in Blanchot, 1988:xxiii.
177 Hannafin, 2004:12.
178 However, as said earlier, Agamben's messianism is the end of the law without significance, and for that, more radical than Derridean (e.g., Derrida, 1992b) messianism.
179 Kafka, 1961.
180 Agamben, 1999a:174; also 1998.
181 Agamben, 1999a:171.
182 Agamben, 1999b:153.
183 Agamben, 1999b:159; for Agamben, messianism is the perfect overturning of an inadequate state of the law – a proposition that explicitly departs from the Derridean dealings with messianism, e.g. 1992a, as the event that can erupt at any moment, and as such cannot be placed on a horizon of waiting. Still, the difference can be assuaged in a combination of the Derridean fractal diffusion of Agamben's reversal, which, however (and *contra* Agamben), cannot wait.

of its internal rupture.[184] In the end, the image of the messianic is the banquet of the righteous, presented not with human heads, but with those of animals. In this rabbinic image of the last day, Agamben does not read the eventual materialisation of continuum between human and natural ('the righteous with animal heads . . . do not represent a new declension of the man-animal relation'),[185] but rather the need for rupture, for the revelation (of the ignorance) of the space of ignorance which allows one to step outside the self: they are 'saved precisely in their being unsavable'.[186] The messianic reveals itself in the fractal reiteration of the continuum/rupture, hooking on its manifestations and bringing its absence to the fore.

Thus, community and rights, both manifestations of the form, are the ways in which law and the city deal with their fears. Neither the law nor the city can deal with *whatever* communities of no presence and identity. On Agamben's Tiananmen Square a cathedral of absence is erected, which marks the limits of the system.[187] Fear of the law is also law's fear, and for this law hides away from absence. Likewise for the city. But it is a short leap from absence to presence of absence. Both systems have been exposing themselves to the offsprings of their couplings – rights is only one of those. Rights such as Blanchot's right to insubordination[188] and to insufficiency,[189] Luhmann's historical understanding of the right to resist,[190] Lefebvre's right to the city,[191] and 'empty' environmental rights as described above, come into play and interrupt the works of the system in ways in which the system has invited, and which remain invisible except as continuous interruption and interrupted continuum of its operations. This is the function of rights: not the interruption of functional dedifferentiation, nor human mobility, but the perpetuation of continuum/rupture in a mutually undercutting way that distils itself in the unutterable paradoxicality of rights. Rights are to be found after the withdrawal – in the absence of marking, and the revelation of the absence of the other side, the side of the environment that remains always excluded but always within. They can be neither of the solitude nor of the community but of the absence of community as felt in its withdrawal – and as such, these rights can only be described in their fluctuation between this and that side, expressing the indescribable and placing an operational absence in the absent

184 See Douzinas, 2000a and 200b, for a similar conclusion through an emphatically non-autopoietic terminology.
185 Agamben, 2004:92.
186 Agamben, 2004:92.
187 Agamben, 1993.
188 Blanchot, 1993:11.
189 Blanchot, 1988:8, my emphasis: 'insufficiency cannot be derived from a model of sufficiency. It is not looking for what may put an end to it, but *for the excess of lack* that grows ever deeper even as it fills itself up.'
190 Luhmann, 2004:360*ff*.
191 Lefebvre, 1996; see also Butler, 2006.

centre of the system: Is this law? The city? Politics? Is this representing a right to a certain environment? A certain right of the environment? A continuum between 'us' and 'not-us'? The rupture of the 'not'? All these are there, a grand mêlée in the blind spot of presence. This is not just a reinstatement of the paradox of rights, but mainly a way of wedging between the two sides of the paradox a space of absence – even better, a movement of withdrawal – populated by the ignorance as to who populates it, which, while bridging the two sides of the paradox together, also maintains them separate in a gesture of operationality. For this reason, rights are necessary and already here: gestures so small that seemingly achieve nothing, yet overturn everything.

V IN WITHDRAWAL

Is there margin in the system's drift for such a withdrawal? Placing absence so violently in the very absence of centre is tantamount to an expectation of radical systemic self-questioning, so radical that it requires a preparedness as to its imploding consequences. Is it the kind of questioning that a system can do? Luhmann says no. The system pushes the difficult questions away into its environment, deeper into other systems' folds. I say, maybe. A system certainly does push away: such movement has appeared in the context of risk, and will appear even more explicitly in the following chapter, where seeds of self-questioning are being converted into waste. But at the same time, and it *is* at the same time, the system succumbs to its questioning through the very movement of pushing away. Its fear of/desire for community, of the liminality of its absence, is immured within the systemic boundaries, and upsets the system's operations from within. The system cannot get rid of its *whatever* singularity, because if it did, it would have to get rid of its very own inability to describe itself – ultimately, it would have to get rid of itself. A slightly dizzy system is better than a schizophrenic one – but who can tell whether systems really have avoided schizophrenia without instituting a schism?

In their movement away from their environment, systems do get dizzy. There is no community between systems, and this is their community. There is no singularity of the system, and this is the systemic singularity. Systems exchange their re-turn to their horizons, pulling their environments together, not in community but in withdrawal, revealing thus the only markable space in the picture: the departure from their in-between. Systems with human heads allow no resting place except for the very porosity of the boundary, and then only after the departure and only after the departure before that. Who dares to speak for the boundary? Who can turn their backs to the in-between and reveal the unutterable paradox? Not rights. They can talk for no one, and this is a good thing. Rights bring forth the paradox, wedge in full inoperability between the horizons of systems, and reveal the absence of the environment.

But these have been invited by the system. In its blind unworking, the system left the door open, and its self-questioning has made a home in its absent centre. The system deals with it – and it is perhaps time to acknowledge that the system does not deal with it very well, and that probably this is it. *Not dealing with* it very well is better than *dealing* it, passing it over. Or perhaps, the system both deals with it and deals it away, in a paradox that shows the system to be cleverer than what Luhmann seems to think, but less clever than what other systems think.

Chapter 6

Waste

Openness, memory and forgetting

The past is not present. The use of the present tense ('is') when referring to the past is a construction that attests to the constructed nature of the past itself. The past cannot 'be' in the present without being present. If the past is remembered, relived or re-enacted, it is present. If the past is forgotten, it was *past – it no longer* is. *The past can only be of a system and never collective. If it is collective, it is a multiple present construction, or a construction of the present, and never of the past. The past cannot be constructed as past – only as present. And that construction would, once again, be a construction of a system, which couples and co-evolves with other systemic constructions. The problem with the past, however, is not that it is a construction – there is nothing unique to this. The problem is that the past of a system was only of the particular system and no system has access to it, not even the system whose past it was. The fact that the past* is *not and can never* be *means that the system cannot access it. The system expels its past to its systemic environment and refuses to itself any accessibility to it. The expelled past may become part of other systems, not as their past but as their present. As far as the system is concerned, its past is invisible. Except for the past that is, namely the remembered past (which is already present). This remains part of the system and forms the memory of the system, which, in its turn, facilitates systemic cognition.*

The part of the past that is *(memory), is retained in the system in the form of self- and hetero-reference. The part of the past that* was *(forgotten), is thrown back into space, and stands little chance of being reselected by the system, because it has been expelled from the system as unnecessary, inoperative, burdensome, obsolete: in short, waste. In this context, waste goes beyond the non-selected and includes a reference to future avoidance. Waste becomes risk. Waste passes in the atemporality of the systemic environment because this is a normal operation of the system. To discard is part of the system's becoming. Not all of the already selected and used is waste – some is retained in the form of memory – but what is deemed to be waste is already out there, a cleared cache from the system's capacity.*

Waste appears in two forms of interest here: the first is urban waste, which is analysed autopoietically from the legal point of view. Second, the legal waste, or else, the non-selected part of a legal decision, which has been discarded to the systemic environment as obsolete. In both cases, the systems attempt to forget the discarded by consigning it to the environment, thereby alienating it temporally and spatially. While this is a necessary operation of the system's being, I suggest that both systems are too keen on discarding the 'unusable' with a consequent loss of valuable cognitive opportunities. In this chapter, the absent environment appears in yet another guise. Its absence claims an operative position in the definition of the system through the means of cognition, memory, and the concept of becoming. This discussion, however, presupposes an exposition of the autopoietic concept of cognitive openness and systemic evolution, which has been hinted at on several occasions so far and here is the appropriate place to analyse it.

I OPENNESS

Autopoiesis likes to think of closed systems as open. A system is closed in its operations, but open cognitively to other systems. This openness is a necessary precondition for the system's evolution, without which the system would not be able to learn. One has the impression, however, that it is also a concession in view of the explanatory problems of constructivism, an impression strengthened when reading either Luhmann or Teubner. Luhmann, for example, started by reiterating Morin's paradoxical adage 'the open rests on the closed',[1] and proceeded into accepting that there exists what he calls a 'legal periphery', which is the contact zone between legal and other communications – an obvious concession to what is generally seen as the implausibility of structural coupling.[2] Likewise the early Teubner espoused an indisputably staunch form of constructivism, which was progressively moderated with the acceptance of a greater, more 'realistic' emphasis on openness.[3] Openness in Teubner acquired historicity: through a sophisticated analytical mechanism, Teubner argues that autopoietic systems are not 'born' but progressively formed.[4] The concept of *hypercycles* describes the process whereby elements of the system – in Teubner's case, the legal system – become independent through self-description and self-constitution, or else, self-observation and self-production. These elements begin to free themselves from social values and acquire a life of their own, a legal life. But the system is not autopoietic until these elements generate further elements through the utilisation of existing elements and processes. Thus, the self-referential cycles are linked together in a self-reproductive hypercycle. Only then can we talk about closure in the autopoietic sense.

1   Morin, *'l'ouvert s'appuie sur le fermé'*, 1986:203.
2   Luhmann, 2004, Ch. 7, section V.
3   Compare Teubner, 1989 and 1993.
4   Teubner, 1993, Chapter 3.

While Teubner's account of hypercycle is convincing and its historical qualities readily appreciated, I find that there is a problem with the existence of the elements of the system before they are organised in self-referential cycles. Teubner says that before the legal system becomes autopoietic, law is socially diffused and identical to general social communications. In other words, it is in the environment of the system-to-be. From this, one can safely assume that these communications were part of some other social system – politics, ethics, family structures – because communications do not circulate freely in the environment: communications cannot be of the environment, they can only be *about* the environment, and necessarily are produced and consumed within systems. In this sense, the idea of 'diffused' law must be purely schematic. Legal communications existed within other systems. But only as each system's own communications. Communications are only of the system in which they reside, and can only be understood by that system. What politics understood as what was later to be recognised as law, has always been law and nothing else. Except that, within the political system, law could never be understood as law – *even when law became a system in its own right*. Politics – or any other system for that matter – continues to perceive legal communications (if at all) as its own communications, regardless of whether these also exist in the legal system or whether they have yet to form a system. If law did not exist as an autopoietic system, legal communications would have been understood as political communications through the political system's self-reference.

For Luhmann, new systems emerge from existing systems by utilising already existing elements of the system.[5] While this proposition is the basis of my suggestion, there is a difference in emphasis: here I put forward an internalisation of evolutionary becoming as seen through an always-already-there being. Such a determination of the system's identity comes from within the system and according to its own temporality: its past starts when the system starts. Before that, only the environment existed in the form of unmarked chaos. As Christodoulidis puts it, 'just as there is no purchase into the unmarked state before the mark, there is no account of the world that precedes the system'.[6] Not for the system at least, because its temporality can only be of the particular system and not of the environment. Teubner's suggestion introduces a historicity that seems to lie purely with the second-order observer and has nothing to do either with the eternal present of Luhmann, or with the intra-systemic time as analysed here.[7]

---

5 Luhmann, 1986a; see also Luhmann, 2004, who suggests the reciprocity of observation as an alternative to Teubner's hypercycles, in a manner more sympathetic to a gradual construction, but without relinquishing the concept of emergence.

6 Christodoulidis, 1998:79.

7 See above, Chapter 4.

Thus, the benefit of acquiescing to systems being systems only when they begin their autopoietic swirls is that, at least on an abstract level, one has to deal neither with a pre-Edenic past, nor with the conventions of second-order observation (to the extent that this is theoretically possible). What is more, seeing everything – including other systems – through the lens of a system, enables a better conceptualisation of the absence of environment within the system as a space of internalised ignorance, which, however, remains foreign to the system. This connects intimately with environmental law's incipient nature. Thus, one could argue that the environmental legal system is not yet autopoietic because it still has not crystallised its elements self-referentially, but relies to a great extent on seemingly external contributions, such as scientific findings, ethical impedimenta and political issues of participation and democracy. Or, one could also argue with Luhmann that environmental law is little more than a cut into pre-existing legal disciplines.[8] The answer to both propositions is simply that, in autopoiesis, a system can only be autopoietic. If it is not autopoietic, it is not a system. If we readily consign the epithet 'environmental' to certain legal communications, it means that, on one hand, these communications are encountered within a system that recognises them as such, and on the other, that these communications embody the external points of reference exactly thus: as points of external observations. It is futile to wonder whether the particular system exists or not: the discourse is lost within its own autopoiesis. The question *exists/does not exist* is a binary code which, once asked, denotes that the concept in question is part of the system: if we ask ourselves whether something is a system or not, then this something is a system, and not only because it perceives itself as such.

There is, however, a difference between incipient and established autopoietic systems. The difference is more quantitative than qualitative and refers to the cognitive capacity of the system. As said, a system is cognitively open to its environment. This openness should be understood as a facet of closure, in that the system invites from its environment only what it can understand as its own. Openness, in other words, perpetuates the self-reproduction of systemic structures as applied to one another, inclusive of uncertainty and limits of ignorance. An established system, namely a system with a more tried-and-tested ability to convert external reference into self-reference, has less use for its environment because of an overpopulation of internal references. An incipient system is more susceptible to external reference for the additional reason that it attempts to formulate its processes in the most complete and self-referential way. Hence, an incipient system will materialise its autopoiesis in a wider 'cognitive domain', that is, the part of a system's horizon whose irritations the system receives and accordingly

8 Luhmann, 1989b.

evolves cognitively;[9] while a better established system already includes a greater portion of its environment within its structures and acts more on self-reference. At the same time, although the space of ignorance for an established system may be larger than that of an incipient one, since the production of knowledge autopoietically generates more ignorance, an incipient system has not yet adequately developed the mechanisms that accommodate self-questioning through internal checks.[10] This means that environmental law may exhibit a greater cognitive capacity than other, more established legal disciplines to deal with its space of ignorance as rendered visible in the absence of its environment.

Of course, the problem remains of how it is that the system performs its cognitive operations. Simply put, systems learn from their environment. A system is cognitively open to its environment, in that it makes its structures available for validity testing or modification. The cognitive exposure of the system is analogous to its observational proactiveness: whatever the system does, it does it to and through itself. As Schütz mentions, the system cannot accomplish any operation outside its boundaries, including on the cognitive level.[11] Learning follows the same reflexive principles of autopoietic reproduction which demand that the cognitive process is a process of superimposition of cognitive layers that project one another in a spiral of self-production: 'learn from experience' takes on an entirely different dimension, for even experience should be understood as the environmental perturbation that instigates in the system the production of another layer of cognitive modification or confirmation of its structures.

Biological autopoiesis suggests that 'learning as a process consists in the transformation through experience of the behavior of an organism in a manner that is directly or indirectly subservient to the maintenance of its basic circularity'.[12] Since the basic circularity of an autopoietic organism is that of being and becoming, learning is inseparable from living: 'living systems are cognitive systems, and living as a process is a process of cognition.'[13] Referring to law, Luhmann accepts the link between normative (being) and

---

9 Maturana and Varela read 'cognitive domain' somewhat differently to each other. The former, as the space of systemic irritations as seen by the observer (1972:10); the latter, as 'the domain of all the descriptions which [the system] can possibly make' (1972:119). The present text obviously veers towards the latter understanding, in view of the possibility of the systemic point of view as the viewpoint of 'reality'. 'Cognitive domain' brings to mind Bourdieu's (1977:95) habitus, or 'a system of lasting and transposable dispositions which, integrating past experiences, functions at every moment as a matrix of perception, appreciations and actions and makes possible the achievement of infinitely diversified tasks'.

10 What Luhmann, 2004, calls redundancy and variation, see below.

11 Schütz, 2000:134.

12 Maturana and Varela, 1972:37.

13 Maturana and Varela, 1972:13; note that in biological autopoiesis, all living systems are autopoietic systems.

cognitive (becoming) operations, but he also differentiates them on the basis of their function. Thus, 'the norm quality serves the autopoiesis of the system, its self-continuation in difference to its environment. The cognitive quality serves the coordination of this process with the system's environment.'[14] The differentiation echoes the distinction made earlier between observations and constitutive operations.[15] Although complementary, the two have different functions for the system. The former provides for the cognitive openness of the system as an avenue of external reference; the latter provides for the identity constitution of the system in the form of internal cohesion by dint of external differentiation. The two interpenetrate and presuppose each other: 'self-reference implies external reference, and vice versa.'[16] Indeed, one of the most noteworthy features of autopoiesis is the indissociability between being and becoming. Any attempt to separate the production from the consumption of meaning is futile: 'the being and the doing of an autopoietic unity are inseparable.'[17] Remarkably, the two operations can be functionally different only because of their indivisibility.

Memory exemplifies the relation between being and becoming. It links what has happened in the past with how it is remembered in the present. Memory is the bridge between cognitive openness ('learn from experience') and operational closure ('learn from *my* experience'). Mnemonic science tells us that the operation of memory is not situated in the past, but in the present: 'remembrance is always now.'[18] In so doing, it connects a systemic becoming – for becoming is not only future but also past – with a systemic being – which invariably lies in the present. What is inscribed in memory – if this metaphor is still allowed – is, on the one hand, the link between past and present by dint of the processual ability of the system to reconstruct its past behaviour in the present, and on the other, the link between the present and the future in the form of an expectation of unperturbed repetition unless something else occurs that would interrupt it. Constructivism teaches us that cognition is *computation of computation of . . .*, or else, a never-ending recursive process of computation.[19] Accordingly, memory must be the speed of computation, or else, the facility with which the system reconstructs its former behaviour when confronted with similar situations to the ones it has faced in the past. Such a computational description of memory is obviously rooted in Aristotle's acts of recollection, according to which, when we remember, we undergo successive earlier changes until we encounter the one after which the change in

---

14 Luhmann, 1988a:20, footnote omitted. For normative and cognitive expectations see above, Chapter 1.
15 Chapter 2, section I.
16 Luhmann, 2004:87, 1993b:52.
17 Maturana and Varela, 1972:49.
18 Steiner, 1975:134.
19 von Foerster, 1984:296.

question usually occurs.[20] Aristotelian theories have had a clear impact on network mnemonic theories, which posit the existence of nodes that connect concepts.[21] But the stauncher expression of computational memory comes from constructivism and biological determinism as seen through autopoiesis. King and Schütz refer to memory as the regular checks for internal consistency performed by the system itself.[22] But memory can also provide for changes, thereby expanding itself and the system. Memory is not always consistent: it includes contradictions and divergences. In such a context, one can talk about memory as *representation*, for memory cannot deliver 'truth', but simply descriptions of it.[23] Such a margin of uncertainty offers the necessary room for impromptu acrobatics, thus enabling the system to deal with novelties intelligently. For, intelligence is essentially the ability to perform new combinations of existing notions. Memory, just as learning, is an operation of selection, which can be conceived as difference, a coupling of two sides: an internal side that links habitually, and manages to link with a velocity analogous to the recurrence of the event, and an external side that links intelligently, in a velocity analogous to the procedural expertise of the system.[24]

Systemic intelligence is tested through asymmetries. Indeed, learning is advanced with asymmetrical positioning of known positions. Explaining how the legal system learns, Jacobson writes: 'cognitive openness introduces asymmetries into the legal system. New cases present new problems of norm-application, hence norm-formation.'[25] A system in perfect accord with its environment, that is, a system free from environmental perturbations, finds itself in a point-for-point mirroring of complexity with its environment. This is not only impossible, since, axiomatically, the environment is more complex than any system,[26] but also undesirable, since the system would never learn. The introduction of asymmetry in the form of environmental complexity is what obliges the system to learn to adjust to new forms of perturbations aided by memory and intelligence. And intelligence is inseparable from the environment, because it is there where the locus of search is. According to

20 Aristotle, 1972.
21 Probably the best introduction to associative and network memory would be Anderson and Bower, 1983.
22 King and Schütz, 1994:140, n.3.
23 Radstone, 2000.
24 Contemporary mnemonic studies offer the analogy of internet engines: 'No single individual can successfully navigate the vast spaces of the internet's database. Instead, intelligent agents navigate and search that information space for their human users, using techniques modelled on the associative processes of human memory' (Locke, 2000:30). In this case, the intelligent agents are less intelligent than the human users: the former can operate only habitually, whereas the latter both habitually and intelligently.
25 Jacobson, 1989:1673.
26 Luhmann, 1995a.

Baecker's comments on Luhmann, 'intelligence starts where an entity is able to take its own lack of knowledge into account and to search for the knowledge it lacks in other entities which presumably are in a better position to bring forth the knowledge sought'.[27] To put it differently, the selection of environmental irritations relevant to the system is a reiteration of the distinction between self- and hetero-reference, with the added space of invited ignorance, in the manner of structural coupling as coupling of environments. The exposition of blind spot and (illusion of) systemic coherence result in the cognitive environment's seeming bifurcation into domesticated and absent. Since both are invited and internalised by the system, thereby converted into the system's cognitive domain, the system learns from the presence of the former and the absence of the latter – and, as I show shortly, waste is a way in which absence is explored.

A final point has to be made before these preliminary remarks are applied to the concept of waste. It may seem from the above that the system selects freely among the environmental perturbations the ones of relevance to its structures. This is only partly true: the freedom of the system is defined by its horizon, which is consequently determined by the system's structural determination.[28] Indeed, 'for any autopoietic system its cognitive domain is necessarily relative to the particular way in which its autopoiesis is realized'.[29] Of course, structural determination is not only phylogenetic, but is modified in time according to the couplings into which the system proceeds. In other words, systemic memory is partly responsible for the selections of the system, hence the difference between established and incipient systems. Memory, in this context, represents the aggregation of procedures selected to be retained by the system. The rest – that is, the non-selected – is discarded into the environment and stands a fair chance never to be selected again, because of the need to *forget*, which is concomitant to the act of remembering. Forgetting is always and irretrievably in the past and, once again, manifests its presence in absence. In a way, the forgotten lies in a slice of the environment as alienated from the system as possible, in a veritable wasteland, conceptually and geographically distanced from the system that needs to forget: this is the way in which this chapter conceptualises the absent environment within the system, whose presence of absence has been forgotten as part of a safe systemic strategy of minimum self-questioning.

While I return to the notion of forgetting below in the context of urban and legal waste, for the present, I feel the need to clarify that the above distanciation of the forgotten in an inaccessible part of the environment is purely a metaphor in terms of autopoietic theory as presented here, which

27 Baecker, 2001:62.
28 See Chapter 2, section VI.
29 Maturana and Varela, 1972:119.

cannot accept any environmental gradation. This *contra* Luhmann, who accepts notions, such as 'periphery' and 'centre' that divide the environment of the legal system into zones of contact and spheres of immediate influence.[30] In the instant interpretation of autopoietic theory, the environment can only be divided once, as selected (already system) and contingent. Further divisions are also possible, but only if a second point of view is introduced, such as a second-order observer who will be able to suppose not only relations, but also quantifications of potentiality of relations. Thus, while a real environment would be nothing but the system, and a utopian environment would be the space outside the system, a 'real' environment is the surprising presence of environmental absence.[31] In that sense, environment (in the 'real' sense, which is the espoused sense here) is not divided into domesticated and absent, but is constituted on the very balance between the two. At present, an explanation for the low potentiality of selection of the forgotten is adequately offered by the role of memory as a guide to present selections. Once again, this does not mean that the forgotten can never be selected. Indeed, this partly is the purpose of this chapter: to show how the forgotten in the form of waste can be reselected by the system. Thus, the systemic wasteland is revisited as the locus of cognitive opportunities through its very absence, its having been forgotten. This is another way in which the absent environment marks its presence of absence within the system.

II WASTE

Waste is what is discarded during and after an operation. Autopoietic operations always involve selections, and, in that sense, waste is the non-selected part of a selection. This is not all that there is to waste, though. The non-selected is contingent, hence it is still within the cognitive domain of the system and can always be selected later on.[32] Waste, as said earlier, does not form part of the cognitive domain of the system as modified by the system's evolution and memory. Waste is labelled by the system as undesired and stands little chance of being considered for selection. Waste is what is discarded.[33] In short, waste is submitted to temporal, geographical and conceptual distanciation from the system. Systems deal differently with their

---

30 Luhmann, 2004 and 1989b.
31 Chapter 2, section II.
32 Although operations take time, which means that, strictly speaking, the non-selected can never be selected in the future. In that sense, waste is indeed the past of the system. Still, the position (as the other of the selected) is contingent.
33 See the art. 1, Directive 91/156/EEC, definition of waste: 'waste shall mean any substance or object in the categories set out in Annex I which the holder discards or intends or is required to discard.'

waste, but they all have this in common: they all produce waste and it is to their immediate benefit to be rid of it as timeously and efficiently as possible. The problem with storing waste in the system is that it creates symmetries with which systemic memory cannot deal. To put it differently, waste makes the absence of the environment visible within the system and brings the system face to face with its limitations in a radical way, connected to its own internal operations.

A further problem with waste comes precisely from its distanciation. By making it scarce, the system positions its waste in a cognitively inaccessible part of its environment. Waste, of course, is being amassed there, usually as part of different systems, which do not perceive it as waste but as their own operation. In the meantime, the memory of the initial system is evolving according to the retained selections. When the irritations from the environment become impossible to ignore, in other words, when waste forces itself into the cognitive domain of the system, the system is unable to cope relying only on its memory and learning abilities, because memory is of the system and, as such, it retains only the selected. The system will have to learn anew how to cope with its discarded non-selections. This does not necessarily mean that the system will not manage. The effect may indeed be the disintegration of the system, but this would be improbable. A far more common way of dealing with it is through structural redefinitions that alter the system while maintaining its autopoiesis.

Faithful to Luhmannian definitional methods, I define waste as the value on the antipodes of *usable*. The binarism can be tangibly explained in the urban context. Urban waste is what is not usable by the urban system; hence it is discarded. The city assigns its waste to its hinterlands, pushes it out of its geography and denies any intimacy with it – or, at least, this is what it tries to do. Urban waste is not assigned to urban memory, for memory is present and waste is by definition past. The city tries to forget its waste and carry on becoming only by retaining a small portion of it as an *aide mémoire*. This urban memory converts waste immediately from past to present, from memory to being: in other words, the memory of the system crosses the binary border and converts waste into usable, through processes such as reusing and recycling. What has not been selected by the system as usable is discarded geographically and denied any memory – that is, any possibility to be present and within our cognitive domain, for, according to Maturana, '[t]he physical space defined as the space in which living systems exist . . . is epistemologically singular because it defines the operational boundaries of our cognitive domain'.[34] Spatial dislocation is a certain way of denuding something of any memory it may carry: the decontextualisation of the object ('Who are you? I know you from *somewhere* but I cannot *place* you!') apart, movement leaves

34 Maturana, 1999:162.

memory behind. Thus Gaston Bachelard: 'memories are immobile, solid only if they are spatialised.'[35] Bachelard famously sees home as the prime locus of memory, where intimacy exists in mnemonic propinquity and immobility. Space is widely recognised as the residence of memories, more so than time, as Pierre Nora's legendary *Les Lieux de Mémoire* painstakingly tries to show, where from a politicised and largely national historicity we are eased into a past remembered as place.[36]

So, the further one transfers waste, the more one denudes it of its memory – and memory here can be read intentionally as the link between the body and the space around it.[37] The further afield waste is from the body, the less remembered it is. Merleau-Ponty's take on memory shows how memory resides both in the 'body-subject' and the space around it as container of objects and processes to be remembered.[38] Bergson defines memory as 'the state of our body', explaining that it is through our bodies that memory becomes the 'progression from the past to the present'.[39] As Casey appositely writes, 'in the actions of the customary body, we observe the continuance of time in place – a continuance that connotes not merely maintenance but active incorporation. In this way the past becomes our true present; it loses its identity as a separate past (a past of another time and place) through its precipitation into the present of bodily behavior, which enacts the past rather than picturing it. And this presentiment of the past is nowhere more active or more evident than in bodily memory of place.'[40] If the body is taken away from its intentional mnemonic object, any memory will be much harder to evoke. This means neither that our cities are clean, nor that we cannot remember the bin in our kitchen. It simply means that in the autopoietic production and consumption of meaning between the body and the city, as it stands at present, there is no place for waste. Waste *was*, and if it *is*, it is part of another system, geographically elsewhere, and it is no longer waste but communication of another system.

If the Bachelardian home is the locus of immobile memories, then contemporary cities must be the locus of forgetting in all its fleetingness and transience. Forgetting is an operation that enables the system to evolve cognitively by discarding the non-selected. Frow sums it up eloquently: '. . . the past is a function of the system: rather than having a meaning and a truth determined once and for all by its status as event, its meaning and its truth are constituted retroactively and repeatedly. Data are not stored in

---

35 Bachelard, 1989:28.
36 Nora, 1984–1993.
37 Casey, 1987:48.
38 Merleau-Ponty, 1962:130.
39 Bergson, 1988:239–40.
40 Casey, 1987:194.

already constituted places but are arranged and rearranged at every point in time. Forgetting is thus an integral principle of this model, since the activity of compulsive interpretation that organizes it involves at once selection and rejection.'[41] Cities reject more because they select more. Their presentocracy means that they forget – they assign everything to achronicity – because of the facility of selection. In his work on the fetishisation of commodities, Baudrillard describes cities as the space of 'limitless promotion of needs',[42] where 'the dictatorship of fashion' demands an excessive production of present and encourages instant forgetting. In fact, the whole orientation of the urban system is towards the production – but not the consumption – of waste. The production of waste rather than goods is the main indicator of abundance, affluence and social pre-eminence: 'waste even appears ultimately as the essential function, the extra degree of expenditure, superfluity, the ritual uselessness of "expenditure for nothing" becoming the site of production of values, difference and meanings on both the individual and the social level.'[43] The urban system forgets as soon as it consumes, spits out before it chews, throws its objects into the past before they are even aired in the present: '[w]hat is produced today is not produced for its use-value or its possible durability, but rather with an *eye to its death*.'[44] The city operates in a perpetual state of dissatisfaction, where only the gadget after the one there is now will do – and gadgets oblige with their in-built obsolescence, and perpetuate a state of conservative affluence, without memory and without cognitive opportunities. The city selects to reject, and its selections are constantly fed by the desire to forget. Overproduction of waste creates an imbalance in the basic circularity of autopoiesis by favouring a hydrocephalous becoming that grows to the detriment of being. The circularity is frozen by an overabundance of selections for the sake of rejection.

III WASTE LAW

Waste is one of the most complicated areas of environmental law that best exemplify the complexity of coupling between law and the city. The two systems embark upon a cognitive trip of mutual observation and mnemonic building. Second-order observations of the particular coupling will probably refer to the common purpose of reserving for the environment a greater role in the mnemonic and cognitive development of the systems. This manifests itself through the concerted efforts of both systems to amplify their memory

41 Frow, 1997:229.
42 Baudrillard, 1998:65.
43 Baudrillard, 1998:43.
44 Baudrillard, 1998:46, original emphasis.

by retaining more waste within the urban system through the legislative tools of reuse and recycle, and to make sure that the discarded waste is dealt with in such a manner that the system will be able to cope with it in the future. From the point of view of the systems, of course, the same findings are expressed differently. There can be no talk about 'purpose' or 'efforts', and any commonality has to be seen internally. For the urban system, irritations from its invitation to legal conditioning result in a redefinition of the city's role as producer and consumer of waste, which involves the introduction of new spatialities (urban hinterlands, contaminated land, location of waste treatment plants) and temporalities (waste excess as risk). For the legal system, the main sources of irritation are, first, the urban excess in waste production, and second, the urban excess in waste expulsion. These are translated as challenges to the legal system's structures that have to be adjusted in order for the system to deal with irritations. In this way, the system learns. The main areas of adjustment are the legal definition of waste, the roles of waste minimisation, recycling and reusing waste, and the continuing responsibility of the waste producer. These areas will be examined below, but before that, a recapitulation of the cognitive process with regard to waste may be helpful to understand the ways of the law.

Systems learn from environmental perturbations, or asymmetries between themselves and their environment, that come about through observations and couplings with other systems. Couplings (in their double sense of structural and environmental) contain regularities, asymmetries, and unmanageable spaces of uncertainty, and as such do not only result in confirmation and reinforcement of structures, but also systemic structural modification and potential malfunctioning. These have a direct impact on the system's memory. Memory can no longer be understood as a depository of notions, but as the speed of processual computations that always takes place in the present. But not everything is retained in memory. Forgetting is also an important cognitive tool, for otherwise the system would not be able to evolve. Forgetting is always in the past. Waste is what has been forgotten and expelled by the system to its past. Waste cannot easily be retrieved by memory because it lies outside the present cognitive domain of the system. Waste is always *was*, and if it *is*, then it is not waste but *usable*. Usable belongs to the system and can be evoked by memory. The boundary between waste and usable is permeable, although the geographical and conceptual distanciation between system and waste makes the boundary less yielding than that of a normal binarism.

Environmental law attempts to grapple with the binarism usable/waste in a curiously ineffective way, by simply repeating the paradox that lies in every binarism without selecting. The EU Directive on waste defines waste as any substance that is to be discarded. The term 'discarded' is not explained in the Directive, and the European Commission has noted the 'major terminological disparity' among the interpretations of Member States.[45] Any attempt to deparadoxify the paradox ('waste is what is discarded; what is discarded is waste') is further confounded by the case law of the European Court of Justice, which has interpreted the term 'discarded' as both the disposal of waste and its consignment to a recovery operation.[46] Moreover, there seems to be a presumption in favour of 'waste' when the material is being consigned to a recovery operation, because such an act constitutes evidence of its being discarded – hence, waste![47] The latter shows how impermeable the boundary between usable and waste is: even that which can be used is considered waste.[48] At another point, the European Court of Justice confirmed that recyclable materials are still considered waste, however much valued by their recyclers.[49] A final example comes from the US Federal legislation. In the US there is an ongoing debate as to whether municipal wastewater should be called 'sewage sludge' or 'biosolids' – in other words, whether legislation should deal with it as waste or usable.[50] The definitional problem reaches its peak when the material in question can be used as a fuel elsewhere, and indeed it is, but can cease to be called waste only after it has been burnt as fuel.[51]

One would expect the law to deal with the excess in waste production by amplifying the definition of non-waste, thus not allowing materials to eschew the system's memory so easily. The reasons for such an inclusive definition of waste are probably connected to the safety guarantees emanating from the duty of care that accompanies waste.[52] Still, if usable, there is no need for disposal. The problem becomes more complicated when materials that habitually fluctuate between usable/waste are taken into consideration, e.g. waste paper, whose demand may sometimes be sufficiently high to justify paying for it, and at other times sufficiently low to justify paying to dispose of it.[53] The law attempts to stop environmental asymmetries from disturbing it – hence, it

---

45 Commission of the European Communities, 1997:4. Note that the commission has drawn up a list containing materials to be considered waste, but the circularity is repeated since for a substance to be included in the list, it has to satisfy the Directive's definition of waste.
46 See the Opinion of A-G Jacobs in *Tombesi* (Case C–304/94) [1997] CMLR 673; also *Van de Walle and Others* (Case C–1/03) [2004] 7 September 2004, not yet reported, and commentary in McIntyre, 2005.
47 *ARCO Chemie Nederland and Others* (Joined Cases C–418/97 and C–419/97) [2000] ECR I–4475.
48 *Zanetti and Others* (Cases C–206/88 and 207/88) [1990] ECR I–1461, where the court rejected the argument that substances capable of economic re-utilisation should not be considered waste.
49 *Tombesi* (Case C–304/94) [1997] CMLR 673.
50 Goldfarb *et al.*, 1999; Stensvaag, 1994.
51 *Castle Cement v Environment Agency* [2001] 2 CMLR 19.
52 Pocklington, 1997.
53 Laurence, 1999:43.

opts for an unyielding boundary between usable/waste. As a result, the law has to deal with the intersystemic claustrophobia of these values: 'once waste, always waste.'[54] The problem is aggravated by another inflexible binarism established by environmental law: that is between the producer and the receiver of waste. In the example of a receiver of, say, subsoil which has been discarded by a construction company, and who does not discard it but uses it for his own purposes, one would assume that the subsoil is no longer waste but usable. This, however, is not accepted by the law: whether something is waste or not is always defined from the point of view of the person who discards it.[55] Such inflexibility has a positive and a negative consequence. The positive is the producer's duty of care as it appears in UK domestic law, which burdens the producer with the obligation to prevent the treatment, keeping or disposal of the waste by *another* person in a manner likely to cause pollution of the environment or harm to human health.[56] Such a continuing liability indicates an interesting ability of the system *not to forget*. This is not exactly remembering, because, strictly speaking, the material is considered waste, hence out of the system, whereas memory is within the system and always present. The continuing liability of the producer introduces another mnemonic selection: that between the *memory of remembering* and the *memory of forgetting*. The first corresponds to the usual processual memory of the system as it applies to itself; the second is what is retained in the system's memory as negative, in absence, and without possibility of recourse to a substantive re-enaction of the event in the present. It is, at the same time, a desperate attempt on behalf of the system to reinstitute the remnants of its past as present. The system cannot remember what it is that it has forgotten – in this case, the produced waste; it can, however, remember having forgotten. In the coupling between environmental law and the city, the law acquiesces to urban forgetfulness (for the urban, waste is forgotten), but remembers precisely this forgetfulness and produces legal communications about it. The waste may be gone from the urban system but, legally, the original person remains liable for the forgotten.

54 Napier, 1998.
55 As decided in *Long v Brooke* [1980] Crim LR 109.
56 Section 33(1)c, Environmental Protection Act 1990. The same principle is reiterated in the European Union context, according to which the producer is fixed with strict civil liability for damage to the environment caused by the produced waste (Proposal for a Directive on Civil Liability for Damage Caused by Waste, COM (91) 219 final). Interestingly, the producer is not only the person who produces waste, but also anyone who carries out any operation that results in a change in the nature or composition of waste, the importer, the controller, and the person responsible for a waster installation. This is a dubious expression of a well-known environmental principle, according to which the polluter pays the cost of their polluting activity ('the Polluter Pays Principle'), since, strictly speaking, it is questionable whether these persons are the 'polluter' or are seen as the polluter by the legal system (Alder and Wilkinson, 1999:285).

The second consequence of the impermeability between producer and receiver of waste is the usual negative consequence of cognitive stubbornness. By refusing the requisite permeability between the binary values of producer/ receiver, the legal system deprives itself from environmental perturbations that would test its structures for validity. The possibility of fluctuation between the producer of waste and its recipient would entail a temporal redefinition of waste: waste is always *was*, thus of the holder. If it *is*, it is not waste, but usable. In short, a temporal modification of the definitional act is required, namely that the binary value of the material is selected according to the present rather than the past. By selecting according to the past, the material is *de facto* waste; by selecting according to the present, the material may have changed into usable. The law finds this fluctuation difficult to accommodate for two reasons. First, because permeability seemingly questions the legal ability to deliver certainty ('what is waste, is waste and cannot be usable at the same time'), disregarding the fact that the illusion of certainty is detrimental to the cognitive openness of the system towards its environment, because of the accompanying otiosity of structures. Hence, the system misses an opportunity to render the absence of its environment visible, for fear of overexposure of its structural certainty. Second, the coupling between the two systems can only with great difficulty accommodate the, admittedly considerable, synchronicity that such a temporal modification would require. What is present for the law, is past for the urban system, with waste constantly changing hands and temporalities.

Still, the system does engage with absence. Its best attempt can be seen in its mnemonic abilities to re-enact the past as present. Memory is the opposite of forgetting, and although a great deal of waste produce is expelled and forgotten by the urban system while facilitated by law's broad definition of waste, some materials escape their fate as waste and are selected by the system by way of reuse or recycling. The processual nature of memory means that, given certain conditions, the system can perform the passage from waste to usable more readily. In memory, temporalities merge, since memory is always a present re-enactment of the past. Of course, the memory of the system is selective – since forgetting is also operationally necessary – and only operates along the already existing mnemonic structures, with all the usual implications for its cognitive openness as a system. In other words, memory evolves on the same lines as any other system, which includes mutual observations and couplings between systemic memories. Accordingly, the mnemonic mutuality between law and the city enables the structures of recycling to perpetuate, without, however, encouraging any significant expansion, unless a redefinition of legal structures enables urban structures to follow. This is where a redefinition of coupling between systemic memories would include the coupling of forgetting, as the mnemonic equivalent of environmental convergence. This would not equate to a total inclusion, but a memory of excluding as a mnemonic tool for future selections between inclusion and exclusion.

As far as the law is concerned, it is significant that recyclable matter is still considered waste for the purposes of the EU Directive on waste,[57] since, as already mentioned, the term 'to discard' also includes consignment for recovery. It is there that the memory of the system comes in and re-selects the expelled waste by crossing the binarism usable/waste. However, the basic feature of recycling is what has been hailed as its 'symbolic role in beginning to change the nature of western societies and the culture of consumerism'.[58] Indeed, while actual recycling is at disproportionately low levels, when compared to waste disposal,[59] its representation of the crossability of the binary border has acquired symbolic dimensions. This becomes particularly relevant in view of the relevance recycling has for packaging waste.[60] Packaging waste apparently contributes 50 per cent of the volume of the total household waste in western societies.[61] It seems that controlling packaging waste through recycling is tantamount to a battle of symbols: affluence versus nostalgia, excess versus utility. Recycling is the counterpart of the Baudrillardian irony, which promotes waste's positive function as the main indicator of social well-being. To the above, recycling's ability to fuse temporalities must be added, for recycling presentifies the past *and* encompasses the future. The latter not only because of its reliance on memory, but also in that any decision on recycling strategy takes into consideration, however inadvertently, the interests of future generations.[62] Recycling emphasises the role of memory as the systemic operation that brings together the temporalities of the system under the security of the notion of repetition. Because of its symbolic rather than actual importance, recycling can afford to ignore all the familiar doubts about its economic and environmental appropriateness,[63] and embrace the elasticity between usable/waste as the way 'to change the nature of western societies'.

Of course, the best way to deal with waste is by not producing it. This is far from everyday practice, however, since the amount of municipal waste is currently growing at around 3 per cent a year.[64] What is more, waste

---

57 Directive 91/156/EEC.
58 Gandy, 1994:1.
59 Haughton and Hunter, 1994.
60 To wit the EC Directive 94/62/EEC on Packaging and Packaging Waste, which provides for 50 to 65 per cent of the weight of packaging waste to be recovered.
61 Gandy, 1994:25.
62 Alder and Wilkinson, 1999:308. Of course, the introduction of intergenerational equity brings along further prioritisations of needs along the lines of renewable and non-renewable sources, which simplify any recycling decisions.
63 Alder and Wilkinson, 1999.
64 Tromans, 2001:258. Still, attempts to reduce waste can be witnessed in several countries: Koppen, 1994, reports on the three-pronged waste reduction strategy which has been launched in the Netherlands as an informal 'covenant'. Koppen's article is part of the book *Ecological Responsibility*, edited by Teubner *et al.*, 1994, which involves an autopoietic take on self-organisation of the industry in the broader context of legal pluralism.

prevention does not even have the symbolic value of recycling, despite being the first in a list of five priorities introduced by the Community Strategy on Waste Management.[65] The other priorities include recycling, optimisation of final disposal, and regulation of waste transport. In practice, the focus of attention has always been the control of disposal activities.[66] For the urban system, the idea of waste disposal is linked to geographical removal, especially in view of the public outcry a waste treatment plant causes if set in an urban area.[67] The suggested response to local reaction is to improve the public image of waste planning authorities,[68] hence encouraging couplings between politics and locality. The law attempts to deal with geographical distanciation with the planning law principles of regional self-sufficiency and proximity.[69] The urban system is correctly seen by the law as a topology that includes its urban hinterlands in the form of continuum/rupture. In reducing them to the city's surrounding geographical region, the law adjusts to the irritations by maintaining its internal interpretation of the differentiation between rural and urban planning – 'differentiation' in the sense of both continuum and rupture. In this differentiation, the law emphasises the boundary between the rural and the urban, whereas the urban system discounts the difference, perceiving it solely as an ever-expanding and naturally available urban geography – thus, a continuum ruptured by its unidirectionality. It is questionable whether this legal differentiation succeeds in being observed by the urban system,[70] especially in view of the political resonance of urban planning issues. It would seem that what is needed is an irritation of sufficient gravity to unsettle some or all conscious and social systems, thereby causing a chain irritation which will put both mnemonic and cognitive abilities of the system to the test. However, such spectacular events are rare and not always effective – already occurred ecological catastrophes have not enabled spectacular structural re-evaluation.

Fortunately, cognitive evolution does not only come along the lines of paradigmatic catastrophes. Systems continue to learn from asymmetries between themselves and their environment, that is from surprising differences between self- and external reference, or indeed from unsolvable inconsistencies within which threaten with the bringing forth of unutterable paradoxes.

65 As quoted in Laurence, 1999:6.
66 Laurence, 1999; Tromans, 2001; Abbot, 2000.
67 Haughton and Hunter, 1994.
68 As suggested in the Government's Waste Strategy for England and Wales 2000, commented on in Tromans, 2001.
69 See *UK Planning and Waste Management*, Planning Policy Guidance Note 10, para. 6, October 2000, as commented on in Tromans, 2001.
70 See the 'name and shame' tool introduced in the EU Sixth Environmental Action Programme, which has already been applied to the breaches of the legal requirements for treatment of urban wastewater discharges in 37 European Union cities. See 'Cities Named and Shamed over Effluent Discharges', in 74 *EU Focus*, 13–14, 2001.

This 'threat' is not the beginning of a paradigm shift – it has always been around, conceptualised as external reference and thus uttered and domesticated. However, the point where waste, risk and concepts of ecological justice have started clogging up systemic operations is already in the past, already *was*, poisoning the memory of the system, and as such haunting the system from within. The system is facing it – not much is there to be done. The system questions itself through its very bodies of appearance, indeed through its absent spaces within. In its couplings with other systems, the system exposes its unutterances, and cognitive abilities that foray into the halo around the open, undomesticated horizon are being developed. In such couplings of environments, where memories of forgetting are being reciprocally (non-)observed while remaining undomesticated, each system resorts to its structural drift and deals with absence within its abilities. What takes time is the resemiologisation of systemic abilities. Thus, the urban system does not ignore the principle of proximity – it couldn't. But it prioritises it (amongst other legal irritations, or other systems' irritations), or responds to it with a temporal differentiation, or responds, but not particularly zealously. These differences in structural modifications do not depend on power scales (e.g., economic considerations are more powerful than legal), but solely on the code of the system and its compatibility with the irritations. The code, though, is not the only tool the system has in its cognitive evolution. Luhmann has offered a couple more cognitive tools in his specific application to the legal system, and this is what I propose to look at in the following section. The discussion will be conducted on the more generic level of the state of law, which, however, is of particular relevance to environmental law because of its incipient, constantly changing behaviour, which, in its turn, leads to an overproduction of legal 'waste'.

IV LAW'S WASTE

Law is as wasteful as the city. Its obligation to select is part of its self-description. It is bound by expectations and self-expectations, and cannot refuse to apply itself to a case. People expect the law to be present when called for; the law expects itself to learn from its selections. These expectations form a series of inescapable moments in which the balance of lawful/unlawful is determined. There is no alternative outside the box: what we do not know about Schrödinger's cat is only analogous to what the cat does not know about itself. The law can only come out of the box once the selection is made, and then only to fall into another selection in another box. As Schütz remarks, 'as opposed to his Roman forebears, the modern judge who declares *non liquet* breaks the law'.[71] The law cannot abandon itself. Thus, the law is

---

71 Schütz, 2000:113. See also Luhmann, 2004.

obliged to select.[72] Selections are asymmetrical, in that they prioritise one branch of the bifurcation over the other. The non-selected is abandoned as inconsistent with the system's memory. It does not enter the memory of the system, but is expelled to the system's environment with a slim chance of being reselected, because of its conceptual distanciation from the system. The non-selected part of the selection is law's waste.

Law's waste follows the usual fate of any systemic waste: it is forgotten. The system forgets, not only because otherwise its selectional mechanisms become clogged, but also because it feels the need to be rid of the uncertainty inherent in every selection. This is the case especially when the domain is the environment, where both external data and internal consideration of the data change constantly. However, a clarification is needed: strictly speaking, what becomes waste is not the non-selected part of the legal decision as such, but *its possibility of having been selected*. Thus, the contingency of the non-selected is reduced because it is discarded in such a way that it no longer fits the binary code of the system. When the legal decision finds an act unlawful, the non-selected 'lawful' is no longer the object of the question 'lawful/unlawful'. It is past the moment of selection and has come out the other end. The law moves on, plunging into further moments of selection, while the possibility of the act being 'lawful' is left behind, because it can no longer involve the system.[73] Law becomes disengaged from this conceptual distanciation – which is the equivalent of the urban system's geographical distanciation. Such a distanciation exemplifies the legal system's expectations of certainty. As Christodoulidis writes, '[t]he law provides a constancy peculiar to it alone. This is due to the function law has in society of stabilising expectations, of controlling normativity, of guaranteeing that its expectations will not be discredited if disappointed, that alter is bound by the legal norm and will bear the consequences if she defies it.'[74] Such a responsibility is what keeps law apace with its self-description: law binds expectations temporally and guarantees certainty infinitely. The law feels the need to forget for fear that the non-selected will return and compromise the certainty that law expects of itself. Law expels its waste and forgets about it, because law's waste questions law, shakes its very processes, threatens with disintegration. As a

---

72 The law cannot use the 'rejection value', that is a third value that negates the binary code as the basis of choice (Luhmann, 1986b). Law cannot describe itself with such a margin because of fear of disintegration. The only tool available to law is what Luhmann, 1997a, calls *steering*, which is explained as the reduction of difference.

73 In the event that the *same* act returns to the system asking once again the question lawful/unlawful, the system will accept it only if it is a *different* act, that is, a different uncertainty as it appears at a different moment, for the law does not engage with identical questions, but only with variations of the first question and only in some cases where the judicial hierarchy allows it. If a *similar* act irritates the system, the system will respond by employing its memory and deliver the outcome of its selection accordingly.

74 Christodoulidis, 1998:107.

matter of survival, the legal system casts its waste away from its cognitive domain, deep into an unthematised horizon that succeeds selection, reducing uncertainty by retaining only certainty.

But things can never be so clear-cut. The moment of selection is always a step away from the unutterable paradox and deeper into some utterability. In law, such utterabilities are usually administered by courts – and this is where Luhmann returns to the discussion and confers to the courts a 'central' position in the management of law, in contradistinction to other forms of legal communication, such as legislature that correspond to the 'periphery' of the system.[75] Luhmann was quick, though, to dispel any impression of hierarchy between centre and periphery, since hierarchy is a misinterpretation of autopoietic circularity.[76] It is simply that the courts are the elements of the legal system that deparadoxify law's fundamental paradoxes, not least because their operation itself is based on a constant need for deparadoxification – that of the legal obligation to decide;[77] but also because courts are the guards of systemic memory, or else the facilitators of its evocation. The courts bring together the past, present and future by reproducing past and already assimilated computations in order to decide a case in the present, and by taking into consideration the future consequences of their decision.[78] The courts – just as memory – are also intentional in that they link an internal with an external element. This is better observed through the concept of *programme*: the courts are the main applicators of the computation between an *if* and a *then*, or else a fact and a legal rule.[79] The conditionality of the application (*if . . . then . . .*) is what a programme is. This conditionality renders the programme more flexible than the binary code and facilitates a more practice-informed application of the code. When a court employs a programme, it links the internal self-referentiality of the code with the avenue of external connection as provided by a programme.

The same internal/external differentiation can be witnessed in another Luhmannian binarism, that between *redundancy* and *variation*.[80] Variation is the systemic accommodation of surprise, whereas redundancy is akin to the memory of the system. While there is little doubt that the system accommodates surprises according to its memory by producing communications, variation triggers not only evocation, but also new combinations – in other words, intelligence. Redundancy, on the other hand, is the process of banalisation of external perturbations that takes place in strict accordance with the

75 Luhmann, 2004.
76 Luhmann, 2004:277, 1993b:302.
77 Luhmann, 2004:292, 1993b:320.
78 Luhmann, 2004:282, 1993b:308.
79 Luhmann, 1989b; 1998a; 2004.
80 Luhmann, 1995b.

system's memory. The system is expected to balance both functions without compromising either its ability of cognitive openness to innovations, or its structural unity. In balancing, the system takes into consideration two kinds of consequences: the intra-systemic consequences which refer to future legal decisions, and the external consequences, or the effects a decision has on the legal environment.[81] It is not as if redundancy and variation have respectively internal and external consequences, or that the legal system can consciously select which mode of reaction it will employ. Rather, the connection is one of contingently balancing one binarism against the other without any prioritisation.

Luhmann observes with consternation the law's historical overgrowth of external over internal reference.[82] I do not share the same concerns. External reference is the way in which the law exposes itself to asymmetries and deals with them by evoking its memory and employing its intelligence. The quantitative difference between external and internal reference is always within the system, a predetermined way of dealing with its environment. The problem is precisely related to the completeness of this dealing. What Luhmann does not take into consideration is, as it were, the external side of the external reference. External reference, as repeatedly argued throughout here, is but the systemic parallelism with its domesticated environment. Both internal and external reference are the ways in which the system checks itself for consistency. The more consistent the system finds itself, the more complacent it becomes.

Legal complacency comes under criticism from Ladeur, who finds Luhmann 'a bit too old fashioned' with regard to the 'iffishness' (if/then) of Luhmannian legal operations.[83] Ladeur comes from a network-based theoretical background, to which he links autopoiesis. The offspring is a pluralistic, anti-universalist, self-organisational theory that questions some of the more conservative aspects of Luhmannian autopoiesis, offering in their stead a market-oriented approach.[84] Ladeur's overarching suggestion is that, since state law in postmodernity is no longer what it used to be, and its substitution with organisations is progressive but inevitable, the only viable way out is the self-organisation of these organisations. While reservations as to the compatibility of the above with ecological considerations, particularly with regard to the environmental record of most transnational organisations, would not go amiss, the fact remains that state law is no longer the only source of law.[85] Accepting other sources of legality is simply a natural consequence

---

81 Luhmann, 1995b:294.
82 See the methodological discussion in Luhmann, 1995b:296–7. See also Baxter, 1998:2030.
83 Ladeur, 1999:26.
84 See also Ladeur and Prelle, 2001, for a specifically environmental development. The possibility of combining autopoiesis and network theory has also been suggested by Murphy, 1997.
85 Teubner, 1997.

of the lack of a vantage point in the autopoietic state of things: 'a distributed order of decisions as a compensation for the impossibility of the position of the ideal observer of society.'[86] This is the antidote to Luhmann's insistence on functional differentiation, and the alternative to what risks becoming an all-describing epistemological reality, rather than a contingent 'reality'. As Ladeur puts it, 'I think autopoietic theory should not be regarded – as Luhmann regards it – as the one and only scientific construction of modern law, but rather as a concept for the description of post-modern law because it is more open to plural concepts of law.'[87]

Along with the fragmentation of legal theory, Ladeur also points to the surrounding conditions of uncertainty, variably attributed to technology, science, demise of linear causality, society itself and so on,[88] that contribute to the 'postmodernisation' of society and increasingly question the foundations of law.[89] In view of institutional (internal) fragmentation, on one hand, and constant conditions of (external) uncertainty, on the other, the legal system is faced with the need to have recourse to a greater openness towards its environment than previously, through what Ladeur calls *modelling*: '[a] strategic model formation under conditions of uncertainty: as a standard of correctness, only the "viable", self-confirming practice can be valid, which has to be explicitly attuned to learning, in a provisional rationality of experimenting with relationing-possibilities.'[90] In other words, no separation between internal and external reference, and certainly no prioritisation of the former can be accepted under the present conditions of indeterminacy. What in this extract Ladeur calls 'modelling', he later calls *proceduralisation*, or else, 'experimental relational rationality',[91] which involves internal learning on the basis of partial, fragmentary knowledge. As he writes characteristically, the legal system '. . . should be more oriented at establishing new kinds of procedural rules stressing flexibility, innovation, experimentation, incentives for long-term horizons of decision-making . . . within a strategy oriented at managing the unexpected through the generation of learning capabilities'.[92] This occurs not with pre-established norms and goals but with patterns, which emanate not from 'the illusionary goal of attaining justice in concrete cases',[93] but from experimentation and internalisation of uncertainty.

86 Ladeur, 1999:22.
87 Ladeur, 1999:33.
88 Ladeur, 1995b;1999.
89 Ladeur 1999, 1995a, 1995b, criticises both the need for legitimisation as explained by Habermas, and the idea of a universalistic, positive law as propagated by Luhmann.
90 Ladeur, 1995a:30, footnotes omitted.
91 Ladeur, 1999:19.
92 Ladeur, 1995b:14.
93 Ladeur, 1999:19.

Ladeur's critique of autopoiesis is valid and nuanced. Even so, it is not difficult to see that what Ladeur calls 'managing the unexpected' is already covered, albeit less adventurously, by Luhmann's external reference. Ladeur's suggestion can be taken as a shift of emphasis on the basis of quantitative priority between internal and external reference – and as such, they are necessary indeed. Where I find that things become more complex is when a conceptualisation of the absence of certainty/uncertainty binarism enters the discussion. Ladeur's suggestion is full of motion and patterns, but I am not certain that it allows for the loss of such patterns, indeed for the unpatternable. The pluralistic criticism could be usefully placed alongside another criticism, this time from deconstruction and Drucilla Cornell's reading of Luhmann (and further, William Rasch's evocative reading of Cornell).[94] For Cornell, there is a space within the system which represents the absence of exteriority. This space entails a responsibility that cannot be discounted via external reference, but only through an act of remembering the system's exclusions.[95] In his turn, Rasch launches a sympathetic critique on Cornell's transcendentalism, which peaks in the responsibility of the system's anthropomorphised interior to speak for the excluded.[96]

Before connecting this to waste, a technical understanding of the matter is attempted. While waste is expelled from the cognitive domain of the system, it offers at the same time a circumference of presence within the system in the manner of the memory of forgetting. This involves a progressive temporal inversion from past-to-present and present-to-future, to present-to-past. Such an inversion requires the level of flexibility readily offered by a system's programmes rather than its code, since the former can accommodate the experimental aspect of the inversion. The only problem may be that the flexibility of a programme is temporally orientated towards the future, without any possibility of reapplication to the past or differentiated application in the present, because such an application would seemingly abandon the characteristic intentionality between internal and external reference. But programmes can be programmed according to the code.[97] Here, programmes will not be required to deviate from the code but only to reinforce its contingency through a reconsidered fusion of temporalities. As I have already mentioned, if present for the law is the lair of certainty, then the forgotten past is the wasteland of ignorance. The past, however, can present itself as a cognitive opportunity for mnemonic re-evaluation: the law can look back and face its waste (in its absence). The cognitive consequences of such turning back can be as freezing as they are bracing: looking Eurydice in the eyes may

94 Cornell, 1992b; Rasch, 2000.
95 Cornell, 1992b:147*ff*.
96 Rasch, 2000:86*ff*.
97 According to Luhmann, 1989b, programmes are given conditions for the suitability of the selection of operations.

make the law come to terms with its limitations and its self-assured capacity for certainty. Indeed, the legal system's boundaries shrink when its waste is brought back to its cognitive domain, and this only leads to a re-evaluation of the expectations and the expectations of expectations of the system. By inverting the temporal orientation of legal selections, thereby reintroducing past non-selections (that are excluded from memory and the system as waste) as absences of selections, the legal system is evoking the waste of selections past, a past that 'lingers like a bad conscience in the shadows of the world from which it has been banned'.[98] This is not an ethical call (it is not understood as such), but simply the reintroduction of the other side of redundancy. If redundancy is the operation of the system with which stabilised grounds and patterns are reactivated[99] – i.e., the memory of remembering – then its flipside will not be variety as external reference, but the *memory of forgetting* as the negative internal formula, which is activated whenever an irritation, similar (but contrary) to the one that led to the construction of memory, appears in the cognitive domain of the system. Since memory is processual, memory of forgetting will be, not what the selection was or was not, but how the selection did not take place. It is not the wasted that returns, but the act of throwing out the waste – the very line of forgetting. To recall Pascal, 'I wanted to write what I forgot, but instead I write that I have forgotten.'[100] The memory of forgetting reactivates the contingency of the alternatives and the paths that lead to their rejection. In so doing, it instils doubt in the form of regular revisiting of choices made and done with.

Such an inversion, which approaches self-questioning, takes place mainly in the courts, since they are considered the 'centre' of legal communications, obliged by the law to apply its code.[101] Of course, the inversion questions the centrality of the courts itself, since a process of demythologisation of the judicial ability to deliver certainty will be in operation. Regardless of their described position, however, it is debatable whether courts can take up the double challenge of maintaining both their centrality, and the self-description of the system as the locus of irreversible certainty. Forms that can accommodate revisiting of the process can be norms constructed in such a way that they include uncertainty in the form of 'fuzzy' conditionality.[102] Or even forms that are not quite norms, but informal arrangements, 'soft' legal hybrids, mediation, self-organising mandates, economic instruments of

---

98 Rasch, 2000:86.
99 Luhmann, 2004.
100 Quoted in Barthes, 1975:23.
101 It is interesting to note that the centrality of the judicial features only in Luhmann's later descriptions of the legal system (2004). Ladeur, 1999:27, explains that the shift is due to 'the weakening of the rule-based universalistic paradigm of law'.
102 Flournoy, 1994; Tarlock, 1994.

internalisation and so on, all of which have already been suggested in the ambit of environmental law.[103] In its most basal form, memory of forgetting can be envisaged as the second-order observer dealing with dissenting judgments, which, although deprived of precedent power, offer the possibility of a post-decisional self-check. Thus, it would seem that courts may have a greater role in the inversion; but at the end, it is the legal system as a whole that deals with its spaces of absence – and this is indeed the case with the way environmental law progressively describes itself: more and more as a *sui generis* branch of law, more and more in need of continuous self-checks, more and more at risk of revealing its paradox. Then, a paradox of the legal system, perilously close to the unutterable, may be how utopia is maintained without law.[104]

The memory of forgetting points, not to the excluded, but to the process of non-selection. Remembering how waste becomes waste is a perambulation on the recesses of the past, where inversions of temporality can only happen around absences of memory. *Eppur si muove*: according to Serres, the trial of Galileo institutes a space of non-law within the law, 'a reservoir of references, of things to refer to outside the law'.[105] And the court is the place to remind the law of this: the court 'as a place of contact or recording, a sieve, ticket window, or semi-conductor between the two worlds . . . the inside and outside of societies, the worldly world and the other'.[106] The case may be, however, that the court is more of a ticket window than a semi-conductor. It is a one-way porosity and its external check is always internalised. Waste and its memory of forgetting, however, tickle a different legal angle, a different movement. The memory of forgetting relies on a withdrawal from space, from the environment, but without reintroducing the thing back into the system. The memory of forgetting, not unlike the other movements previously commented in this book, is not about presence but about absence, and only then – *only after* – can it be of presence. Remembering, for the systems in question, is the institution of the absence of the thing past, the selection never-made, and the internalisation of the presence of such absence. The form memory/forgetting neither enables continuum, nor emphasises rupture, but does both without doing either. The space of non-law is not accessible even to such a flexible memory. It remains ignorance, limits, limitations, and does all that only through its boundary, situated inside the system. The thing remains absent – the environment can never appear as external reference, as patternable field of action, or even as a ghost. It is only its boundaries that put in a spectral appearance, and if the system manages to escape its fear

103 Teubner, 1996; Shelton, 2000.
104 Ellickson, 1991.
105 Serres, 1995:83.
106 Serres, 1995:63.

of/desire for it, then it can convert it into the external side of the external reference, a temporal inversion, an utterable paradox. But only after.

V IN MEMORY

The systemic environment cannot be remembered: if it is, then it will be part of the system, an operation of the present. The forgotten past of the system is expelled into the environment, pushed as far away as possible in the horizon, and deprived of any immediate possibility of realisation of its inherent contingency. Waste must be put out of sight because it irritates the system. It is then, perhaps, only a matter of measurement how much waste is excessive and when a system should reconsider its waste production. Arguably, the clearest indication is when the waste produced begins to irritate the system itself as a recurrent present problem. Then, the system is forced to take its waste into consideration and accommodate it through its normal operations. But at points like this, the system may also consider the possibility of reducing waste altogether rather than simply dealing with it afterwards. In other words, prevention rather than rectification.

At such points, the absence of the environment is traced within the system. Absence is populated by those parts of the system that the system itself no longer desires. This goes beyond the tools of external reference and variation, although it does have a cognitive effect. Systemic openness is transferred inside and in relation to the absolute closure of the absent environment. And from this, the system learns – that is, not from its ability to internalise irritations, but from its very inability. The system learns by exposing itself to its limitations, thus constructing the external side of its external reference and progressively accommodating its very own inability, its space of non-system within its boundaries. Several modes have been suggested for this, but the point is that there is no alternative for the system than to carry on dealing with its ignorance without, however, pat(t)ernalising it.

As one way among others, the memory of forgetting has directional resonance with the withdrawal witnessed in the previous chapters. What is introduced in the system in this way is not the thing but the way, the system's mnemonic tracing of the other side of its selection, the *how* of the unmarking. The thing cannot be marked – time has passed, justice cannot be applied case-by-case – but the system will be marked with its unmarking, and its selections will be informed both positively (as self-checks for mnemonic consistency) and negatively (as self-checks for leithic consistency). Thus, the system carries on throwing out, wasting, rejecting. But in its jolting production, it consumes its trace of throwing, of crossing – and this is maintained in the system, in promise of the form that cannot be uttered.

Opening

. . . the city must both amass around itself to possess identity, and escape itself in order for its unbearable identity to be liberated.
Barber, 1995:26

I

Having the environment as a focus in a way that avoids ecocentrism is difficult. Conceptualising the environment as absent while avoiding anthropocentrism is more difficult. But talking about the thing that negates communication because of its absence is perhaps impossible. The paradox comes to the rescue, only to disappear behind a veil of utterabilities. Still, is not admission of ignorance itself knowledge? How much is it allowed to know about the ecology of ignorance in order to declare it absent? This is the form continuum/rupture at work: rupture visibilises the continuum through its own continuity, and vice versa. One can only carry on venturing innocently, always unaware of the rupture round the corner, for otherwise the continuum cannot even begin.

The city is what there is, environmental law is how it is, and their environment is the impossibility of knowing how it is not. Blind spots offer suspicions of absence, but never absence itself. Absence can be beatified only through the halo of its disappearance, and then iconoclastically subjugated to its presence, but only after. Eyes on withdrawal: this is not passivity, even at its most radical, nor resistance, even at its mildest. Withdrawal is the present movement in whose folds past and future, recursivity and oscillation, simultaneity and endocausality, absence and its presence are interdigitated. No hierarchy, no priority, no centre. Withdrawal itself is to be located in the movement deeper into the excentricity of absence.

Ecology returns home, to the *oikos* of the *logos*, and interrupts logos by leaving oikos behind. Ecology can only do what it can do: depart and talk about nostalgia, the return, the talking itself. Flights of utterabilities circle its absence, making the system, not so much construct the facts of ecology (it has been doing that anyway), but construct its presence around its invitation to the absence of its environment. Inside is what outside is, an unutterable absence, whose boundaries carve deep into the system's inner recesses and make their presence felt as the limitations of the very system.

Autopoiesis not quite on its head: from system to the absent environment and then back to the system in its traumatised emptiness. And back out again. All these movements are already in the theory, already happening. This text simply reminds the theory of its movement. It takes it for a walk and brings it back home. Safety redefined.

II

Abstraction in general, and autopoietic abstraction in particular, allows for a transdisciplinary mode of description that encourages a convergence of legal and geographical issues. Through abstraction, independence from social situations is achieved, which is particularly welcomed here in view of the generality rather than specificity of the city as the geographical canvas on which legal operations are observed. But as often happens with theories of great sophistication, autopoiesis, with its terminology, potential applications and concurrent epistemology, operates very technically on an abstract level, so much so that its technicality becomes anti-abstract. This creates a need for a transposition of both high abstraction and high concreteness of technicalities to any practical application the specific theory may be subjected to, which, in its turn, makes any application, from the abstraction of transdisciplinarity to the tangibility of the operability of one discipline, particularly difficult and often inelegant.[1] In any combination of theoretical and practical analyses, one of the two usually sheds its plumes and proceeds to a kind of simplification for the sake of fitting. And while authors such as Teubner,[2] Christodoulidis,[3] Rogowski,[4] King and Thornhill,[5] or Murphy[6] have presented us with admirably coherent autopoietic descriptions of the practicalities of law, a discrepancy remains between Luhmann's complexity of descriptions and the selections that the authors had to make in order to present a coherent application. This does not indicate that something is wrong with subsequent theoretical applications. On the contrary, the discrepancy should be attributed to the fact that even when Luhmann gratifies the

---

1   Luhmann himself, 1993b:24, admits that his theory is not supposed to guide practice, although the emphasis falls on 'to guide'.
2   Especially Teubner, 1993, 1996.
3   Christodoulidis, 1998.
4   Rogowski, 1994.
5   King, 1994; King and Piper, 1990; King and Thornhill, 2003.
6   Murphy, 2001.

reader with practical examples, he remains consistently abstract and technical, arguably because he is as interested in developing the all-informing
aspect of autopoiesis as he is in proving points about its empirical relevance.
In the case of subsequent literature, however, a major consideration has been
to use autopoiesis as a support for particular theses or, to the extent that
every thesis develops its pedestal theory, to develop theoretical aspects of
autopoiesis as contextualised within a chosen system.

In full awareness that a similar inelegance may be witnessed in the present
text, I can only admit to the sin of selection on aspects of both theory and
practice. In my attempt to apply autopoiesis to law, I have made some choices
which seemed better equipped for such an application. Thus, the fact that I
have chosen to describe the precautionary principle, risk, environmental
rights, and waste, and leave out practices, such as environmental impact
assessment, air pollution, land use, and whatever else one cares to remember,
dispels any expectation of exhaustiveness or conclusiveness. It only reveals
the intent behind these selections to combine the legal and the urban with
autopoiesis, and to present not a totalising theory, but an indicative description of such applications as seen by the instant observer. There is little doubt
that other selections could have been made that would adequately fit the
autopoietic remarks that preceded these applications, and vice versa. On
the other hand, I cannot profess that all selections would have accorded with
the theory as presented here. The not exceptionally dishonest solution in this
case would have been to modify the theory to accommodate the practice,
since one cannot change empirical reality, only its theoretical description . . .

III

Autopoiesis does not attempt to change anything, only describe. This is true
to the extent that autopoiesis does not assume the proactive panoply of other,
notably critical, theoretical strands. In the words of Tim Murphy, autopoiesis
'does not as such challenge the existing organisation of society, the distribution of power and resources, or the rightness and authority of the law and
legal institutions. It does challenge the meaningfulness of how the forms and
functions of these arrangements are often understood. It does not take up
cudgels on behalf of the oppressed; it is not a critical-emancipatory theory.'[7]
This does not mean that autopoiesis is devoid of controversy. Far from it, its
controversy is internalised in the form of paradox, and any potential for
change represents another instance in the successful failure of deparadoxification. As Teubner explains, '[d]e-paradoxification means to invent new distinctions which do not deny the paradox but displace it temporarily, and thus

---

7   Murphy, 1994a:248.

relieve it of its paralysing power.'[8] But these distinctions are always internal, just as the paradoxes they try to inertisize. Change, future, novelty, and probabilities are already included in autopoiesis in the form of distinctions. What differs is their descriptions, which then operate as a trampoline on which contingency performs its acrobatics.

The present text does not advocate change in the critical sense, since it is fully aware of its structural limitations as merely another piece of theory trying to communicate its findings to itself. Nevertheless, it is more adventurous than a typical autopoietic description. While it remains a description and avoids prescription, it often espouses a certain optimism and ambition which, although typically autopoietic, departs from the Luhmannian dogma of 'dreamless' pragmatism and proceeds to what has been described in the second chapter as 'reality's' proclivity to utopia as a realisable project. The opening to the movements of systems has informed concepts, such as unutterance, withdrawal, community, continuum/rupture, and has allowed a waft of absence to breathe in the systemic closure of autopoiesis. This does not mean that this description 'takes up cudgels' on behalf of the unrepresented. All it does is reshuffle the basic autopoietic distinctions and immanently suggest a different description of the above distinctions that seems to be justified, even for Luhmann, in view of the peculiarity of ecological considerations. Autopoiesis is an appropriate tool to describe such redefinitions, because it ensures that they remain 'realistic', humbly systemic and devoid of the bravado of sweeping social reforms – in short, well into their systemic limitations. The question of course remains, whether this sort of description, indeed this absence, is operationalised by the system – more specifically, whether environmental law and the city are systemically capable of dealing with their absent environments within. Autopoietically, the answer can only be this: if it is within, then it is dealt with. And in practice the answer would be similar: both environmental law and the city are still around, more or less dealing with the issues. The question, then, is not so much whether the system will keep on functioning, but rather, whether it can differentiate itself in a way that would surprise its self-description. One cannot do a great deal more than carry on describing; but describe the absence one can.

IV

The way the term 'environment' has been employed in the book is indicative of the level of abstraction on which the text mainly unfolds. Its use without a qualifying adjective encourages its abstract – or ambiguous – connotations. This is not to say that environment is a unifying concept. The environment

8   Teubner, 2001a:32.

institutes difference, and as such one can talk about unity (of difference). But the system does not look for its unity outside. On the contrary, unity is to be found on the movement of withdrawal from absence to the presence of absence.

The environment of the title, the environments of the text, are not the domesticated environments of the systems, the locus of the systemic external reference. The environments of absence are whatever is not included in the system's self-description except as a nagging hole of incommunicability. This 'opening' of the concept of autopoietic environment, not qualified by communications, revealed the space in which ignorance could be accommodated. Within the broader horizon of the system – whatever the system – the environment has been described as the intentional relation of continuum/rupture between the system and its ignorance. As such, the environment is ecological.

Such an absence has only been possible after the discussion on Luhmannian society and 'reality'. The rethinking of autopoietic society meant that any inclusion in the autopoietic environment could no longer be mediated by the maternity of the supra-system, but had to find other ways of rendering its absence visible. Thus, the concept of unutterance emerged as the vehicle of the unutterable paradox of the system, the *memento vanitas* of the system within the system. Unutterance balances on the internal construction of systemic limitations, on the boundary between knowledge and ignorance, revealing the pores that link while separating, presentifying the absence while preserving its inaccessibility. The system, like Michel Serres's judge in Gallileo's case,[9] is shaken up before its continuum/rupture with what has been invited within. The system cannot escape such an invitation – it is the thing that does while being, its very being and becoming, its operation of observation that generates blind spots as loci of questioning. The environment breezes in and institutes a space of non-system within the system.

For reasons peculiar to each system, the environment proceeds from a general to a particular invisibility. For the environmental legal system, this is because of its expectational idiosyncracies. For the urban system, it is on account of its autopoietic construction that consists of episodes of intensification. For both systems, the environment has been described in a relation of continuum/rupture to be located in the movement between ignorance and knowledge. The purpose of the above has been simply to provide the basic theoretical considerations on which the further particularisation of the absence of the environment would take place. On an autopoietic theory level, the environment could be defined as a system-specific negativity. On an ecological level, the protection of nature does not presuppose understanding of natural operations, and certainly not a presumption of unfractured identity between the human and the natural.

9   Serres, 1995.

From then on, the environment has been concretised in its absence through various temporal and situational operations. Thus, nowhere does the difference between the presence of the system and the absence of environment become more obvious than in the future of systems. Starting from the Luhmannian axiom that everything that happens, happens in the present, and that nothing that happens to/in the system can be found anywhere else but the system itself, systemic future is located in the absent environment, as internalised by the system, when the latter deals with future risks. The idea of operational negativity expressed by unutterance finds its exemplification in the description of the precautionary principle as the vertiginously self-referential process of projecting the present into the future, only not to project it, namely to withdraw from it. In all its cognitive impossibility, absence is conceptualised as undomesticated presence by the system, and any scientific doubts as to the repercussions of the projection are muted by the postponement of decision in view of the presumption in favour of environmental protection that precaution calls for. Indeed, it is through the operation of internalisation that the system becomes aware of its own limitations before the uncertainty of the future and the incomprehensibility of natural operations, and chooses, instead of the usual systemic colonisation of the unknown, to deal with uncertainty by not dealing with it, by respecting the fact that it does not know and that until it knows more, there is little point in projecting.

The knowledge of ignorance proceeds to the centre of the system, only to show that in the centre there is absence. Around this absence, community clusters. Community is neither a conscious nor a societal manifestation of becoming. It escapes such categories, principally because of its ability for being internalised as withdrawal. Thus, systems turn to themselves, to their absent centre, for the community of absence. An ecological community, an urban community that internalises its continuum/rupture with its absent environment, locates itself in the placelessness of the re-turn. Without nostalgia, community proceeds by undermining itself and its exclusions, on the knowledge of an exclusion that applies principally to itself. Community excludes community, and on the basis of such an exclusion, community can be ecological. But who speaks for the community – or for whom does the community speak? The answer, of course, involves no speech. Rights left out in the environment, environmental rights that tear themselves on the boundary of their bifurcation are tangible materialisations of unutterance, utterabilities that hail the messianic as it lies next to them, in misrecognition.

In the end, the book has reserved a manifestation of ignorance at its most ruptured continuum with the self. With the concept of waste, the system witnesses a part of itself being expelled and distanced as undesired. The operation is routinely performed in order for the system to evolve mnemonically and cognitively. Forgetting has been described both temporally – as past – and geographically – as the distanciation into the autopoietic environment.

The reversal of this distanciation, but on the basis of a firmly established difference, is the way in which the absent environment makes its last appearance. The environment is now included in its absence in the manner of a reversal of forgetting: not by remembering – thus converting the other into self – but through the memory of forgetting, which retains the selection between memory and forgetfulness (or self and other), but allows the process in which the other has become other to haunt the system. With this, asymmetries are reinstated as cognitive opportunities for the system vis-à-vis its environment, and as the autopoietically viable way for the much sought after unity in difference.

V

The book can be easily accused of being both ecologically pessimistic and autopoietically optimistic. Both stem from the role the environment has assumed. It is ecologically pessimistic because the natural environment remains expelled from the system as the bottomless depository of ignorance and uncertainty. However, this is the autopoietic expression of my misgivings as to the importance the natural environment actually has in the quotidian processes of communication and perception. With this in mind, I was looking for a way to express my worries about the environment without, however, abandoning a pragmatic (or else, pessimistic) viewpoint. Thus, I tried to advocate for a role for the environment that could be sustained as a matter of description without having to rely on prescriptive utopianism. On the other hand, my stance can be described as autopoietically optimistic, because, despite its absence (or precisely because of it), the environment is included in the system both definitionally and operationally. Indeed, in my reading of autopoiesis I attempted to insert what I lack from an ecological point of view: a dosage of utopian idealism. The absence of the environment has been internalised as the embodiment of incommunicability through unutterance. In the ability offered in autopoiesis to escape communication lies the theory's appropriateness as an explanatory tool for the complexity of relations between the human and the natural. The autopoietic capacity for internalisation, although *prima vista* claustrophobic and restrictive, has proven to be a 'realistic' tool with which to address the demythologised role of the natural environment and to offer instead a system-specific (mis)understanding of what the role of the natural environment may actually be. Whether this understanding is of any use to the environment is debatable. But it may be of some use to the system.

Bibliography

Abbot, C., 'Waste Management Licensing: Benefit or Burden?', *Journal of Planning and Environmental Law*, 1003–1010, October 2000

Adorno, T., *Negative Dialectics*, trans. E. Ashton, New York: Seabury, 1973

Adorno, T. and Horkheimer, M., *Dialectic of Enlightenment*, London: Verso, 1997

Agamben, G., *The Coming Community*, trans. M. Hardt, Minneapolis: Minnesota University Press, 1993

Agamben, G., *Homo Sacer: Sovereign Power and Bare Life*, trans. D. Heller-Roazen, Stanford: Stanford University Press, 1998

Agamben, G., 'The Messiah and the Sovereign: The Problem of Law in Walter Benjamin', in *Potentialities: Collected Essays in Philosophy*, ed. and trans. D. Heller-Roazen, Stanford: Stanford University Press, 1999a

Agamben, G., 'Walter Benjamin and the Demonic: Happiness and Historical Redemption' in *Potentialities: Collected Essays in Philosophy*, ed. and trans. D. Heller-Roazen, Stanford: Stanford University Press, 1999b

Agamben, G., *The Open: Man and Animal*, trans. K. Attell, Stanford: Stanford University Press, 2004

Åkerstrøm Andersen, N., *Discursive Analytical Strategies: Understanding Foucault, Koselleck, Laclau, Luhmann*, Bristol: Policy Press, 2003

Alder, J. and Wilkinson, D., *Environmental Law and Ethics*, London: Macmillan, 1999

Allen, J., Massey, D. and Pryke, M. (eds), *Unsettling Cities*, London: Routledge, 1999

Amin, A. and Thrift, N., *Cities: Reimagining the Urban*, Cambridge: Polity Press, 2002

Anchor, R., 'Whose Autopoiesis?', 1 *History and Theory* 39: 107–16, 2000

Andermatt Conley, V., *Rethinking Technologies*, Minneapolis: University of Minnesota Press, 1993

Anderson, B., *Imagined Communities*, London: Verso, 1991

Anderson, J. and Bower, G., *The Architecture of Cognition*, Cambridge, MA: Harvard University Press, 1983

Anderson, M., 'Human Rights Approaches to Environmental Protection: An Overview', in Boyle, A. and Anderson, M. (eds), *Human Rights Approaches to Environmental Protection*, Oxford: Clarendon Press, 1996

Arendt, H., *The Origins of Totalitarianism*, San Diego: Harcourt Brace & Company, 1958

Aristotle, Περι Μνημης [*De Memoria et Reminiscentia*], trans. R. Sorabji, London: Duckworth, 1972

Aristotle, *Metaphysics Z*, trans. H. Tredennick, London: Loeb, 1989

Bachelard, G., *La Poétique de l'Espace*, Paris: Quadrige, 1989

Baecker, D., 'Why Systems?', 18 *Theory, Culture & Society* 1: 59–74, 2001

Balke, F., 'Tristes Tropiques. Systems Theory and the Literary Scene', 8 *Soziale Systeme* 1: 27–37, 2002

Banakar, R., *Merging Law and Sociology: Beyond the Dichotomies in Socio-Legal Research*, Berlin: Galda & Wilch, 2003

Bankowski, Z., 'How Does It Feel to be Alone? The Person in the Sight of Autopoiesis', in Nelken, D. (ed.), *Law as Communication*, Aldershot: Dartmouth, 1996

Barber, S., *Fragments of the European City*, London; Reaktion Books, 1995

Barthes, R., *Roland Barthes par Roland Barthes*, Paris: Seuil, 1975

Bartoloni, P., 'The Stanza of the Self: on Agamben's Potentiality', *Contretemps* 5: 8–15, 2004

Bateson, G., *Mind and Nature: A Necessary Unity*, New York: Bantam, 1988

Batty, M. and Longley, P., *Fractal Chaos*, London: Academic Press, 1994

Baudelaire, C., *The Painter of Modern Life and Other Essays*, trans. J. Mayne, Da Capo Paperback, n.d.

Baudrillard, J., *America*, trans. C. Turner, London: Verso, 1988

Baudrillard, J., *The Transparency of Evil: Essays on Extreme Phenomena*, trans. J. Benedict, London: Verso, 1993

Baudrillard, J., *The Consumer Society: Myths and Structures*, London: Sage, 1998

Baudrillard, J., *Simulacra and Simulation*, trans. S. F. Glaser, Michigan: Ann Arbor – The University of Michigan Press, 2000

Bauman, Z., *Globalization: The Human Consequences*, Cambridge: Polity Press, 1998

Bauman, Z., *Community: Seeking Safety in an Insecure World*, Cambridge: Polity Press, 2001

Bauman, Z., *Liquid Love*, Cambridge: Polity, 2003a

Bauman, Z. 'Utopia with no Topos', 16 *History of the Human Sciences* 1: 11–25, 2003b

Bauman, Z., *Living in Utopia*, Public Lecture, LSE (London), 27 October, www.lse.ac.uk/collections/LSEPublicLecturesAndEvents/pdf/20051027-Bauman2.pdf, last accessed 10.11.05, 2005a

Bauman, Z., 'Seeking Shelter in Pandora's Box, or: Fear, Security and the City', 9 *City* 2: 161–8, 2005b

Baxter, H., 'Autopoiesis and the Relative Autonomy of Law', 19 *Cardozo Law Review* 6: 1987–2090, 1998

Beauregard, R., 'Theorizing the Global-Local Connection', in Knox, P. and Taylor, P. (eds), *World Cities in a World-System*, Cambridge: Cambridge University Press, 1995

Beauregard, R., 'City of Superlatives', 2 *City and Community* 3: 183–99, 2003

Beck, A., 'Is Law an Autopoietic System?', 14 *Oxford Journal of Legal Studies*: 400–18, 1994

Beck, U., *Risk Society: Towards a New Modernity*, London: Sage, 1992

Beck, U., *Ecological Enlightenment: Essays on the Politics of Risk Society*, New Jersey: Humanities Press, 1995

Beck, U., 'World Risk Society as Cosmopolitan Theory? Ecological Questions in a Framework of Manufactured Uncertainties', 13 *Theory, Culture and Society* 4: 1–32, 1996

Beckerman, W. and Pasek, J., *Justice, Posterity, and the Environment*, Oxford: Oxford University Press, 2001

Bell, D., *Husserl*, London: Routledge, 1990
Bell, D. and Haddour, A. (eds), *City Visions*, Harlow: Prentice Hall, 2000
Bell, M., *An Invitation to Environmental Sociology*, California: Pine Forge Press, 1998
Benhabib, S., *Situating the Self: Gender, Community and Postmodernism in Contemporary Ethics*, Cambridge: Polity Press, 1992
Benhabib, S., 'From Identity Politics to Social Feminism: A Plea for the Nineties', in Trend, D. (ed.), *Radical Democracy*, London: Routledge, 1996
Benjamin, W., *Gesammelte Schriften V, Das Passagen-Werk*, Frankfurt: Suhrkamp, 1982
Benjamin, W., *Charles Baudelaire: A Lyric Poet in the Era of High Capitalism*, trans. H. Zohn, London: Verso, 1983
Benjamin, W., *One-Way Street and Other Writings*, London: Verso, 1985
Bennett, M. and Teague, D. (eds), *The Nature of Cities: Ecocriticism and Urban Environments*, Tuscon: University of Arizona Press, 1999
Bergson, H., *An Introduction to Metaphysics*, trans. T. E. Hulme, Indianapolis: Bobbs-Merrill, 1955
Bergson, H., *Matter and Memory*, trans. N. Paul and S. Palmer, New York: Zone Books, 1988
Bernet, R., Kern, I. and Marbach, E., *An Introduction to Husserlian Phenomenology*, Evanston, Illinois: Northwestern University Press, 1993
Bernhard, T., *Wittgenstein's Nephew*, Chicago: University of Chicago Press, 1988
Bertalanffy, von, L., *Robots, Men and Minds*, New York: Braziller, 1967
Bertalanffy, von, L., *General System Theory*, New York: Braziller, 1969
Birnie, P., *Birnie and Boyle on International Law and the Environment*, Oxford: Oxford University Press, 2002
Blanchot, M., *The Space of Literature*, trans. A. Smock, Lincoln, Nebraska: University of Nebraska Press, 1982
Blanchot, M., *The Unavowable Community*, trans. P. Joris, New York: Station Hill Press, 1988
Blanchot, M., 'Pre-texte: Pour l'Amitié', in Mascolo, D. (ed.), *A la Recherche d'un Communisme de Pensée*, Paris: Fourbis, 1993
Blanchot, M., *Friendship*, trans. E. Rottenberg, Stanford: Stanford University Press, 1997
Blomley, N., *Law, Space and the Geographies of Power*, New York: The Guilford Press, 1994
Blomley, N., Delaney, D. and Ford, R. (eds), *The Legal Geographies Reader: Law, Power, and Space*, Malden, MA: Blackwell, 2001
Blowers, A. (ed.), *Planning for a Sustainable Environment*, London: Earthscan, 1993
Blowers, A., Hamnett, C. and Sarre, P. (eds), *The Future of Cities*, London: Hutchinson, 1974
Blowers, A. and Pain, K., 'The Unsustainable City?' in Pile, S. *et al.* (eds), *Unruly Cities? Order/Disorder*, London: Routledge, 1999
Blühdorn, I., *Post-Ecologist Politics: Social Thinking and the Abdication of the Ecologist Paradigm*, London: Routledge, 2000
Bodansky, D., *Proceedings of the American Society of International Law*, American Society of International Law, 1991
Boehmer-Christiansen, S., 'The Precautionary Principle in Germany: Enabling

Government', in O'Riordan, T. and Cameron, J. (eds), *Interpreting the Precautionary Principle*, London: Earthscan, 1994

Bohman, J. and Rehg, W. (eds), *Deliberative Democracy*, Cambridge, MA: MIT Press, 1997

Bookchin, M., *Remaking Society: Pathways to a Green Future*, Boston: South End Press, 1990a

Bookchin, M., *The Philosophy of Social Ecology: Essays on Dialectical Naturalism*, Montreal: Black Rose Books, 1990b

Bourdieu, P., *Outline of a Theory of Practice*, trans. R. Nice, Cambridge: Cambridge University Press, 1977

Boyle, A., 'The Role of International Human Rights Law in the Protection of the Environment', in Boyle, A. and Anderson, M. (eds), *Human Rights Approaches to Environmental Protection*, Oxford: Clarendon Press, 1996

Bowden, P. and Lawrence, J., 'THORP and After: Challenging State Decisions', *European Environmental Law Review*: 251–6, 1994

Brady, K., 'New Convention on Access to Information and Public Participation in Environmental Matters', 28 *EPL*: 69, 1998

Braeckman, A., 'Niklas Luhmann's Systems Theoretical Redescription of the Inclusion/Exclusion Debate', *Philosophy and Social Criticism* 32: 65–88, 2006

Braidotti, R. *Metamorphoses: Towards a Materialist Theory of Becoming*, Cambridge: Polity, 2002

Braidotti, R., *Transpositions: on Nomadic Ethics*, Cambridge: Polity Press, 2006

Briggs, J. and Peat, D., *Turbulent Mirror*, New York: Harper & Row, 1989

Briginshaw, V., 'Dancing Bodies in City Settings: Construction of Spaces and Subjects', in Bell, D. and Haddour, A. (eds), *City Visions*, Harlow: Prentice Hall, 2000

Butler, C., 'Sydney: Aspiration, Asylum and the Denial of the "Right to the City" ', in Philippopoulos-Mihalopoulos, A. (ed.), *Law and the City*, London: Glasshouse Press, 2006

Butler, J., *Gender Trouble: Feminism and the Subversion of Identity*, London: Routledge, 1990

Butler, J., *Bodies that Matter*, London: Routledge, 1993

Buttel, F., 'Classic Theory and Contemporary Environmental Sociology', in Spaargaren, G. *et al.* (eds), *Environment and Global Modernity*, London: Sage, 2000

Cage, J., *Silence*, Middletown: Wesleyan University Press, 1961

Caldwell, L., *Between Two Worlds*, Cambridge: Cambridge University Press, 1990

Calvino, I., *Le Città Invisibili*, Milano: Arnoldo Mondadori, 1993

Calvino, I., *Il Castello dei Destini Incrociati*, Milano: Arnoldo Mondadori, 1994

Cameron, J. and Abouchar, J., 'The Precautionary Principle: A Fundamental Principle of Law and Policy', 14 *Boston College International and Comparative Law Review*: 1–27, 1991

Cameron, J. and Mackenzie, R., 'Access to Environmental Justice and Procedural Rights in International Institutions', in Boyle, A. and Anderson, M. (eds), *Human Rights Approaches to Environmental Protection*, Oxford: Clarendon Press, 1996

Campiglio, L., Pineschi, L., Siniscalco, D., and Treves, T. (eds), *The Environment after Rio*, London: Graham and Trotman, 1994

Caputo, J., *The Prayers and Tears of Jacques Derrida: Religion without Religion*, Bloomington: Indiana University Press, 1997

Carbonneau, T. (ed.), *Lex Mercatoria and Arbitration*, New York: Transnational Juris Publications, 1990

Casey, E., *Remembering: A Phenomenological Study*, Bloomington: Indiana University Press, 1987

Casey, E., 'How to Get from Space to Place in a Fairly Short Stretch of Time: Phenomenological Prologema', in Feld, S. and Baso, K. (eds), *Senses of Place*, Santa Fe: School of American Research Press, 1996

Castells, M., *The Informational City*, Oxford: Basil Blackwell, 1989

Castells, M., *The Rise of the Network Society*, Oxford: Blackwell, 1996

Chambers, I., *Migrancy, Culture, Identity*, London: Routledge, 1994

Chatterjee, B., 'Cyber Cities: Under Construction', in Philippopoulos-Mihalopoulos, A. (ed.), *Law and the City*, London: Glasshouse Press, 2006

Christodoulidis, E., *Law and Reflexive Politics*, Dordrecht: Kluwer Academic, 1998

Clam, J., 'The Specific Autopoiesis of Law' in Pribáñ, J. and Nelken, D. (eds), *Law's New Boundaries: The Consequences of Legal Autopoiesis*, Aldershot: Dartmouth Ashgate, 2001

Clark, N., ' "Botanizing on the Asphalt"? The Complex Life of Cosmopolitan Bodies', 6 *Body & Society* 3–4: 12–33, 2000

Clifton, T., 'The Poetics of Musical Silence', 62 *The Musical Quarterly*: 163–81, 1993

Cohen, J. and Stewart, I., *The Collapse of Chaos: Discovering Simplicity in a Complex World*, New York: Viking, 1994

Commission of the European Communities, *Communication from the Commission to the Council and the European Parliament Concerning the Application of Directives 75/439/EEC, 75/442/EEC, 78/319/EEC and 86/278/EEC on Waste Management*, COM (97) 23 final, 1997

Connelly, J. and Smith, G., *Politics and The Environment: From Theory to Practice*, London: Routledge, 1999

Cooper, D., *Governing out of Order: Space, Law and the Politics of Belonging*, Rivers Oram: New York University, 1998

Cooper, D. and Palmer, J. A. (eds), *The Environment in Question*, London: Routledge, 1992

Cordonier Segger, M-C., Khalfan, A., Gehring, M., and Tooering, M., 'Prospects for Principles of International Sustainable Development Law after the WSSD', 12 *RECIEL* 1: 54–68, 2003

Cornell, D., 'Time, Deconstruction, and the Challenge to Legal Positivism: The Call for Judicial Responsibility', 2 *Yale Journal of Law and Human Sciences* 2: 267–97, 1990

Cornell, D., 'The Relevance of Time to the Relationship between the Philosophy of Limit and Systems Theory', 13 *Cardozo Law Review* 5: 1579–603, 1992a

Cornell, D., *The Philosophy of the Limit*, London: Routledge, 1992b

Cornell, D., *The Imaginary Domain: Abortion, Pornography and Sexual Harassment*. London: Routledge, 1995

Cotterrell, R., 'Sociological Perspectives on Legal Closure', in A. Norrie (ed.), *Closure or Critique: New Directions in Legal Theory*, Edinburgh: Edinburgh University Press, 1993

Cotterrell, R., *Law's Community: Legal Theory in Sociological Perspective*, Oxford: Clarendon Press, 1995

Cotterrell, R., 'The Representation of Law's Autonomy in Autopoiesis Theory', in

Pribáň, J. and Nelken, D. (eds), *Law's New Boundaries: The Consequences of Legal Autopoiesis*, Aldershot: Dartmouth Ashgate, 2001

Coyle, S. and Morrow, K., *The Philosophical Foundations of Environmental Law*, Oxford: Hart, 2004

Crowe, N., *Nature and the Idea of a Man-Made World: An Investigation into the Evolutionary Roots of Form and Order in the Built Environment*, Cambridge, MA: MIT Press, 1997

Cox, K. (ed.), *Spaces of Globalization: Reasserting the Power of the Local*, New York and London: The Guildford Press, 1997

De Certeau, M., *The Practice of Everyday Life*, trans. S. Rendall, Berkeley: University of California Press, 1984

Deggau, H-G., 'The Communicative Autonomy of the Legal System', in Teubner, G. (ed.), *Autopoietic Law: A New Approach to Law and Society*, Berlin: Walter de Gruyter, 1988

Delaney, D., *Law and Nature*, Cambridge: Cambridge University Press, 2003

Deleuze, G. and Guattari, F., *Anti-Oedipus: Capitalism and Schizophrenia*, trans. R. Hurley, M. Seen, H. Lane, New York: Viking Press, 1977

Deleuze, G. and Guattari, F., *A Thousand Plateaus*, trans. B. Massumi, London: Athlone Press, 1987

Derr, T., *Ecology and Human Need*, Philadelphia: Westminster Press, 1975

Derrida, J., *Speech and Phenomena, and Other Essays on Husserl's Theory of Signs*, trans. D. Allison, Evanston: Northwestern University Press, 1973

Derrida, J., *Of Grammatology*, trans. G. C. Spivak, London: John Hopkins University Press, 1976a

Derrida, J., 'Pas', 1 *Gramma: Lire Blanchot* 3–4: 111–215, 1976b

Derrida, J., *Writing and Difference*, trans. A. Bass, Chicago: University of Chicago Press, 1978

Derrida, J., *Dissemination*, trans. B. Johnson, Chicago: University of Chicago Press, 1981

Derrida, J., *Signéponge/Signsponge*, trans. R. Rand, New York: Columbia University Press, 1984

Derrida, J., *Edmund Husserl's 'Origin of Geometry': An Introduction*, trans. J. Leavey, Jr., Lincoln: University of Nebraska Press, 1989

Derrida, J., 'Force of Law: The "Mystical Foundation of Authority"', trans. M. Quaintance, in Cornell, D., Rosenfeld, M. and Gray Carlson, D. (eds), *Deconstruction and the Possibility of Justice*, New York: Routledge, 1992a

Derrida, J., 'Before the Law', trans. A. Ronell and C. Roulton, in Attridge, D. (ed.) *Acts of Literature*, London: Routledge, 1992b

Derrida, J., *Spectres of Marx: The State of the Debt, the Work of Mourning, and the New International*, trans. P. Kamuf, London: Routledge, 1994

Derrida, J., *Adieu*, trans. P. Brault and M. Naas, Stanford: Stanford University Press, 1999

Derrida, J., *On Cosmopolitanism and Forgiveness*, trans. M. Dooley and M. Hughes, London: Routledge, 2001

Derrida, J., *The Politics of Friendship*, trans. G. Collins, London: Verso, 2005

Derrida, J. and Dufourmantelle, A. *Of Hospitality: Anne Dufourmantelle Invites Jacques Derrida to Respond*, trans. R. Bowlby, Stanford: Stanford University Press, 2000

Desgagné, R., 'Integrating Environmental Values into the European Convention on Human Rights', 89 *The American Journal of International Law* 2: 263–94, 1995

Devall, W. and Sessions, G., *Deep Ecology: Living As If Nature Mattered*, Salt Lake City: Gibbs Smith, 1985

Diken, B., 'City of God', Department of Sociology, Lancaster University, 2004, at http://www.comp.lancs.ac.uk/sociology/papers/diken-city-of-god.pdf last accessed on 15.11.05

Douglas, M., *Risk and Blame: Essays in Cultural Theory*, London: Routledge, 1992

Douglas-Scott, S., 'Environmental Rights in the European Union: Participatory Democracy or Democratic Deficit?', in Boyle, A. and Anderson, M. (eds), *Human Rights Approaches to Environmental Protection*, Oxford: Clarendon Press, 1996

Douzinas, C., *The End of Human Rights*, Oxford: Hart, 2000a

Douzinas, C., 'Human Rights and Postmodern Utopia', 2 *Law and Critique* 11: 219–40, 2000b

Douzinas, C., 'Identity, Recognition, Rights or What Can Hegel Teach Us About Human Rights?', 29 *Journal of Law and Society* 3: 379–405, 2002

Douzinas, C. and Gearey, A., *Critical Jurisprudence: The Political Philosophy of Justice*, Oxford: Hart, 2005

Douzinas, C., Goodrich, P. and Hachamovitch, Y. (eds), *Politics, Postmodernity and Critical Legal Studies: The Legality of the Contingent*, London: Routledge, 1994

Downey, J. and McGuigan, J., *Technocities*, London: Sage, 1999

Doxiadis, C. and Papaioannou, J., *Ecumenopolis: The Inevitable City of the Future*, Athens: Athens Centre of Ekistics, 1974

Dryzek, J., 'Political and Ecological Communication', in Dryzek, J. and Schlosberg, D. (eds), *Debating the Earth: the Environmental Politics Reader*, Oxford: Oxford University Press, 1998

Dumont, L., *Homo Hierarchicus: Essai sur le Système des Castes*, Paris: Gallimard, 1966

Dupuy, J-P., 'On the Supposed Closure of Normative Systems', in Teubner, G. (ed.), *Autopoietic Law: A New Approach to Law and Society*, Berlin: Walter de Gruyter, 1988

Eckersley, R., *Environmentalism and Political Theory: Toward an Ecocentric Approach*, London: UCL Press, 1992

Eckersley, R. (ed.), *Markets, the State and the Environment: Towards Integration*, Basingstoke: Macmillan, 1996

Edensor, T., 'Moving Through the City', in Bell, D. and Haddour, A. (eds), *City Visions*, Harlow: Prentice Hall, 2000

Eder, K., *The Social Constructing of Nature*, London: Sage, 1996

Elder, S., 'Do Species and Nature Have Rights? The Wrong Answer to the Right(s) Question', *Osgoode Law Hall Journal* 22: 285–95, 1984

Elin, N. (ed.), *Architecture of Fear*, New York: Princeton Architectural Press, 1997

Ellen, R., *Environment, Subsistence and System: The Ecology of Small-Scale Formations*, Cambridge: Cambridge University Press, 1982

Ellickson, R., *Order Without Law: How Neighbors Settle Disputes*, Cambridge, MA: Harvard University Press, 1991

Elliot, R., *Faking Nature: The Ethics of Environmental Restoration*, London: Routledge, 1997

Elster, J. (ed.), *Deliberative Democracy*, Cambridge: Cambridge University Press, 1998

Elworthy S. and Holder J., *Environmental Protection*, London: Butterworths, 1997

Engel, J.R. and Engel, J.G. (eds), *Ethics of Environment and Development: Global Challenge and the International Response*, London: Belhaven, 1990

Fainstein, S. and Campbell, S. (eds), *Readings in Urban Theory*, Oxford: Blackwell, 1996

Farber, D., 'Environmental Protection as a Learning Experience', 27 *Loyola of Los Angeles Law Review* 3: 791–808, 1994

Farmer, L. and Teubner, G., 'Ecological Self-Organization', in Teubner, G. *et al.* (eds), *Environmental Law and Ecological Responsibility: The Concept and Practice of Ecological Self-Organization*, Chichester: John Wiley & Sons, 1994

Faulks, J. and Rose, L., 'Common Interest Groups and the Enforcement of European Environmental Law', in Somsen, H., *Protecting the European Environment: Enforcing EC Environmental Law*, London: Blackstone, 1996

Feintuck, M., 'Precautionary Maybe, but What's the Principle?', 32 *Journal of Law and Society* 3: 371–98, 2005

Feld, S., *Waterfalls of Song*, in Feld, S. and Baso, K. (eds), *Senses of Place*, Santa Fe: School of American Research Press, 1996

Ferkiss, V., *Nature, Technology and Society: Cultural Roots of the Current Environmental Crisis*, London: Adamantine Press, 1993

Femers, S. and Jungermann, H., 'Eine Systematisierung und Diskussion von Risikomaßen', 1 *Zeitschrift für Umweltkunde*: 59–84, 1992

Fichte, J. G., *The Science of Knowledge*, ed. and trans. P. Heath and J. Lachs, Cambridge: Cambridge University Press, 1982

Fincher, R. and Jacobs, J. (eds), *Cities of Difference*, London: The Guildford Press, 1998

Fisher, C., 'Is the Precautionary Principle Justiciable?', 13 *Journal of Environmental Law* 3: 315–34, 2001

Fitzpatrick, P., 'The Impossibility of Popular Justice', in de Sousa Santos, B. (ed. of volume), State Transformation, Legal Pluralism and Community Justice, 1 *Social and Legal Studies* 2: 199–215, 1992

Fitzpatrick, P., 'Abiding the World: Globalism and the Lex Mercatoria', in Pribáñ, J. and Nelken, D. (eds), *Law's New Boundaries: The Consequences of Legal Autopoiesis*, Aldershot: Dartmouth Ashgate, 2001

Flournoy, A., 'Coping with Complexity', 27 *Loyola of Los Angeles Law Review* 3: 809–24, 1994

Foerster, von, H., *Observing Systems*, California: Intersystems Publications, 1984

Folke, C. and Kaberger, T., *Linking the Natural Environment and the Economy*, Dordrecht: Kluwer Academic, 1992

Foltz, B., *Inhabiting the Earth: Heidegger, Environmental Ethics, and the Metaphysics of Nature*, Atlantic Highlands, NJ: Humanities Press, 1995

Fordham, M. and Cameron, J., 'Judicial Review and Public Interest Groups', 1 *Environmental Judicial Review Bulletin* 5: September 1994

Fox, W., *Towards a Transpersonal Ecology: Developing New Foundations for Environmentalism*, London: Shambhala, 1990

Freestone, D. and Hey, E. (eds), *The Precautionary Principle and International Law: The Challenge of Implementation*, The Hague: Kluwer Law International, 1996

Frisby, D., *The Flâneur in Social Theory*, in Tester, K. (ed.), *The Flâneur*, London: Routledge, 1994

Frow, J., *Time and Commodity Culture: Essays in Cultural Theory and Postmodernity*, Oxford: Clarendon Press, 1997

Fuchs, P., *Die Metapher des Systems*, Weilerswist: Velbrück Verlag, 2001

Gadamer, H-L., *Truth and Method*, ed. and trans. J. Weinsheimer and D. Marshall, London: Sheed and Ward, 1989

Game, A., 'Riding: Embodying the Centaur', 7 *Body & Society* 4: 1–12, 2001

Gandy, M., *Recycling and the Politics of Urban Waste*, London: Earthscan, 1994

Gaon, S., 'The Question of the Community (is) the Community of the Question: A Certain Sociality and Solidarity in the Thought of Jean-Luc Nancy', in Coward, M. (ed.), Special Issue: *Being-with: Jean-Luc Nancy and the Question of Community*, 9 *Journal for Cultural Research* 4: 387–403, 2005

Gardiner, R., 'Between Two Worlds: Humans in Nature and Culture', 12 *Environmental Ethics*: 339–52, 1990

Gelick, J., *Chaos*, London: Sphere, 1988

Geyer, F. and van der Zouwen, J. (eds), *Dependence and Equality: A Systems Approach to the Problems of Mexico and Other Developing Countries*, Oxford: Oxford University Press, 1982

Geyer, F. and van der Zouwen, J. (eds), *Sociocybernetic Paradoxes: Observation, Control and Evolution of Self-Steering Systems*, Beverly Hills: Sage, 1986

Gilbert, A. and Gugler, J., *Cities, Poverty and Development: Urbanization in the Third World*, Oxford: Oxford University Press, 1992

Gillespie, A., *International Environmental Law and Ethics*, Oxford: Oxford University Press, 2002

Gilloch, G., *Myth and Metropolis: Walter Benjamin and the City*, Cambridge: Polity Press, 1996

Gilroy, P., *Between Camps*, London: Allen Lane, 2000

Girardet, H., *The Gaia Atlas of Cities: New Directions for Sustainable Living*, London: Gaia Books Ltd, 1992

Girouard, M., *Cities and People: A Social and Architectural History*, New Haven: Yale University Press, 1985

Goldfarb, W., Krogmann, U. and Hopkins, C., 'Unsafe Sewage Sludge or Beneficial Biosolids?', 26 *Boston College Environmental Affairs* 4: 687–768, 1999

Golding, M. P., 'Obligations to Future Generations', *Monist* 56: 87–107, 1972

Goodrich, P., *Legal Discourse: Studies in Linguistic, Rhetoric and Legal Analysis*, London: Macmillan, 1987

Goodrich, P., *Languages of Law: From Logics of Memory to Nomadic Masks*, London: Weidenfeld and Nicolson, 1990

Goodrich, P., 'Anti-Teubner: Autopoiesis, Paradox, and the Theory of Law', 13 *Social Epistemology* 2: 197–214, 1999

Goodrich, P., 'First We Take Manhattan: Microtopia and Grammatology in Gotham', in Philippopoulos-Mihalopoulos, A. (ed.), *Law and the City*, London: Glasshouse Press, 2006

Goodwin, B. and Taylor, K., *The Politics of Utopia: A Study in Theory and Practice*, London: Hutchinson, 1982

Gottman, J., *Megalopolis*, New York: Free Press, 1961

Grabham, E., 'Taxonomies Of Inequality: Lawyers, Maps, and the Challenge of Hybridity', 15 *Social & Legal Studies* 1: 5–23, 2006

Graham, S., 'Towards Urban Cyberspace Planning: Grounding the Global through

Urban Telematics Policy and Planning', in Downey, J. and McGuigan, J., *Technocities*, London: Sage, 1999

Graham, S. and Marvin, S., *Telecommunications and the City: Electronic Spaces, Urban Places*, London: Routledge, 1996

Grange, J., *The City: An Urban Cosmology*, New York: State University of New York, 1999

Grosz, E., *Space, Time and Perversion*, London: Routledge, 1995

Grosz, E., *Architecture from the Outside: Essays on Virtual and Real Space*, Cambridge, MA: MIT Press, 2001

Grosz, S., 'Access to Environmental Justice in Public Law', in Robinson, D. and Dunkley, J. (eds), *Public Interest Perspectives in Environmental Law*, London: Wiley Chancery, 1995

Gumbrecht, H. U., 'How is our Future Contingent? Reading Luhmann against Luhmann', 18 *Theory, Culture & Society* 1: 49–58, 2001

Gumbrecht, H. U. and Pfeiffer, K. L. (eds), *Materialities of Communication*, trans. W. Whobrey, Stanford, California: Stanford University Press, 1994

Guzzoni, U., 'Nature: A Theme for Finite Philosophical Thinking?', in Langsdorf, L. *et al.* (eds), *Phenomenology, Interpretation, and Community*, New York: State University of New York Press, 1996

Habermas, J., 'Law as a Medium and Law as Institution', in Teubner, G. (ed.), *Dilemmas of Law in the Welfare State*, Berlin: Walter De Gruyter, 1985

Habermas, J., *The Theory of Communicative Action, ii. Lifeworld and System*, Boston: Beacon, 1987

Habermas, J., *Between Facts and Norms: Towards a Discourse Theory of Law and Democracy*, Cambridge: Polity Press, 1996

Haddour, A., 'Citing Difference: Vagrancy, Nomadism and the Site of the Colonial and Post-Colonial', in Bell, D. and Haddour, A. (eds), *City Visions*, Harlow: Prentice Hall, 2000

Hafif, M., 'Silence in Painting: Let Me Count the Ways', in Jarowski, A. (ed.), *Silence: Interdisciplinary Perspectives*, Berlin: Mouton de Gruyter, 1997

Hajer, M., *The Politics of Environmental Discourse: Ecological Modernization and the Regulation of Acid Rain*, Oxford: Oxford University Press, 1995

Hanemann, P., 'Economics and the Preservation of Biodiversity', in Wilson, E. O. (ed.), *Biodiversity*, Washington: National Academy Press, 1988

Hannafin, P., 'The Writer's Refusal and Law's Malady', 31 *Journal of Law and Society* 1: 3–14, 2004

Hannigan, J.A., *Environmental Sociology: A Social Constructionist Perspective*, London: Routledge, 1995

Haraway, D., *The Haraway Reader*, London: Routledge, 2004

Hardin, G., 'The Tragedy of the Commons', in Markandya, A. and Richardson, J. (eds), *The Earthscan Reader in Environmental Economics*, London: Earthscan, 1992

Harding, A., 'Practical Human Rights, NGOs and the Environment in Malaysia', in Boyle, A. and Anderson, M. (eds), *Human Rights Approaches to Environmental Protection*, Oxford: Clarendon Press, 1996

Hardoy, J., Mitlin, D. and Satterthwaite, D., *Environmental Problems in an Urbanizing World: Finding Solutions in Africa, Asia and Latin America*, London: Earthscan, 2001

Harrison, P., 'Niklas Luhmann and the Theory of Social Systems', in Roberts, D.

(ed.), *Reconstructing Theory: Gadamer, Habermas, Luhmann*, Victoria: Melbourne University Press, 1995

Harremoës, P. *et al.* (eds), *The Precautionary Principle in the Twentieth Century: Late Lessons from Early Warnings*, London: Earthscan, 2002

Harvey, D., *The Condition of Postmodernity*, Oxford: Blackwell, 1989

Harvey, D., 'Social Justice, Postmodernism, and the City', in Fainstein, S. and Campbell, S. (eds), *Readings in Urban Theory*, Oxford: Blackwell, 1996a

Harvey, D., *Justice, Nature and the Geography of Difference*, Oxford: Blackwell, 1996b

Harvey, D., 'The Environment of Justice' in Merrifield, A. and Swyngedouw, E. (eds), *The Urbanization of Justice*, London: Lawrence and Wishart, 1996c

Haughton, G. and Hunter, C., *Sustainable Cities*, London: Regional Studies Association, 1994

Hayles, K., 'Making the Cut: The Interplay of Narrative and System, or What Systems Theory Can't See' in Rasch, W. and Wolfe, C. (eds), *Observing Complexity: Systems Theory and Postmodernism*, Minneapolis, MN: University of Minnesota Press, 2000

Heelas, P., Lash, S. and Morris, P. (eds), *Detraditionalization: Critical Reflections on Authority and Identity*, Blackwell: Oxford, 1996

Heidegger, M., 'Building Dwelling Thinking', in *Poetry, Language, Thought*, trans. A. Hofstadter, New York: Harper and Row, 1971

Heidegger, M., *Being and Time*, trans. J. McQuarrie and E. Robinson, Oxford: Basil Blackwell, 1992

Heidegger, M., 'What is Metaphysics?' in *Basic Writings*, ed. and trans. D. Farrell Krell, London: Routledge, 1996

Henry, M., *L'Amour Les Yeux Fermés*, Paris: Jules Tallandier, 1977

Henry, M., 'Quatre Principes de la Phénoménologie', 1 *Revue de Métaphysique et de Morale*: 3–26, 1991

Highet, K., 'The Enigma of Lex Mercatoria', in Carbonneau, T. (ed.), *Lex Mercatoria and Arbitration*, New York: Transnational Juris Publications, 1990

Hinchliffe, S., 'Cities and Natures: Intimate Strangers', in Allen, J. *et al.* (eds), *Unsettling Cities*, London: Routledge, 1999

Hirst, P. and Thompson, G., *Globalisation in Question*. Cambridge: Polity Press, 1996

Hodkova, I., 'Is There a Right to a Healthy Environment in the International Legal Order?', 65 *Connecticut Journal of International Law* 7: 65–80, 1991

Hohmann, H., *Precautionary Legal Duties and Principles of Modern International Environmental Law*, London: Graham & Trotman/Martinus Nijhoff, 1994

Holder, J. and McGillivray, D. (eds), *Locality and Identity: Environmental Issues in Law and Society*, Aldershot: Ashgate, 1999

hooks, b., *Yearning: Race, Gender and Cultural Politics*, Toronto: Between the Lines, 1990

Howells, C., *Derrida: Deconstruction from Phenomenology to Ethics*, Cambridge: Polity Press, 1998

Husserl, E., *Ideen zu einer reiner Phänomenologie und phänomenologischen Philosophie. Zweites Buch: Phänomenologische Untersuchungen zur Konstitution*, M. Biemel (ed.), The Hague: Martinus Nijhoff, 1954

Husserl, E., *The Idea of Phenomenology*, trans. W. Alston and G. Nakhnikian, The Hague: Martinus Nijhoff, 1964

Husserl, E., *Logical Investigations*, trans. J. Finlay, London: Routledge & Kegan Paul, 1970a

Husserl, E., *The Crisis of European Sciences and Transcendental Philosophy*, trans. D. Carr, Evanston: Northwestern University Press, 1970b

Husserl, E., *Cartesian Meditations*, trans. D. Cairns, The Hague: Martin Nijhoff, 1973a

Husserl, E., *Experience and Judgment*, trans. J. S. Churchill and K. Ameriks, Evanston: Northwestern University Press, 1973b

Husserl, E., *Phantasie, Bildewusstein, Erinnerung, 1898–1925*, ed. E. Marbach, HuA XXIII, 1980

Husserl, E., *Ideas Pertaining to a Pure Phenomenology and to a Phenomenological Philosophy, First Book*, trans. F. Kersten, The Hague: Martinus Nijhoff, 1983

Husserl, E., *On the Phenomenology of the Consciousness of Internal Time (1893–1917)*, trans. J. Barnett-Brough, Dordrecht: Kluwer Academic, 1991

Iyer, L., 'Our Responsibility: Blanchot's Communism', *Contretempres* 2: 59–73, 2001

Jacobson, A., 'Autopoietic Law: The New Science of Niklas Luhmann', 87 *Michigan Law Review*: 1647–89, 1989

Jacobson, A., 'The Idea of a Legal Unconscious', 13 *Cardozo Law Review* 5: 1473–505, 1992

James, A., 'An Open or Shut Case? Law as an Autopoietic System', 19 *Journal of Law and Society* 2: 271–83, 1992

Janicaud, D., 'Should a Phenomenologist be Clever?', in Langsdorf, L. *et al.* (eds), *Phenomenology, Interpretation, and Community*, New York: State University of New York Press, 1996

Jarowski, A., 'White and White: Metacommunicative and Metaphorical Silences', in Jarowski, A. (ed.), *Silence: Interdisciplinary Perspectives*, Berlin: Mouton de Gruyter, 1997

Jay, M. (ed.), *An Unmastered Past: The Autobiographical Reflections of Leo Lowenthal*, Berkeley: University of California Press, 1987

Johnson, R., 'Blurred Boundaries: a Double-Voiced Dialogue on Regulatory Regimes and Embodied Space', 9 *Law Text Culture*: 157–78, 2005

Kafka, F. 'Before the Law', in *The Penal Colony: Stories and Short Pieces*, trans. W. and E. Muir, New York: Schocken, 1961

Kahneman, D., Slovic, P. and Tversky, A., *Judgment under Uncertainty: Heuristics and Biases*, Cambridge, MA: MIT Press, 1982

Kauffman, S., *At Home in the Universe: The Search for Laws of Complexity*, Harmondsworth: Viking, 1995

Kavka, G. S. and Warren, V., 'Political Representation for Future Generations', in Elliott, R. and Gare, A. (eds), *Environmental Philosophy*, Milton Keynes: Open University Press, 1983

Keil, R., Bell, D., Penz, P. and Fawcett, L. (eds), *Political Ecology: Global and Local*, London: Routledge, 1998

Kellogg, C., 'Love and Communism: Jean-Luc Nancy's Shattered Community', 16 *Law and Critique*: 339–55, 2005

Kelsen, H., *The Pure Theory of Law*, trans. M. Knight, Berkeley: University of California Press, 1967

King, A. (ed.), *Re-presenting the City: Ethnicity, Capital and Culture in the 21st Century Metropolis*, London: Macmillan, 1996

King, M., 'The Truth about Autopoiesis', 20 *Journal of Law and Society* 2: 218–36, 1993

King, M., 'Children's Rights as Communication: Reflections on Autopoietic Theory and the United Nations Convention', 57 *Modern Law Review* 3: 385–401, 1994

King, M., 'The Construction and Demolition of the Luhmann Heresy', 12 *Law and Critique* 1: 1–32, 2001

King, M., 'An Autopoietic Approach to "Parental Alienation Syndrome"', 13 *The Journal of Forensic Psychiatry* 3: 609–35, 2002

King, M. and Piper, C., *How the Law Thinks about Children*, Aldershot: Gower, 1990

King, M. and Schütz, A., 'The Ambitious Modesty of Niklas Luhmann', 21 *Journal of Law and Society* 3: 261–87, 1994

King, M. and Thornhill, C., *Niklas Luhmann's Theory of Politics and Law*, London: Palgrave Macmillan, 2003

King, M. and Thornhill, C. (eds) *Luhmann on Politics and Law: Critical Appraisals and Applications*, Oxford: Hart, 2005

Kiss, A., 'The Rights and Interests of Future Generations and the Precautionary Principle', in Freestone, D. and Hey, E. (eds), *The Precautionary Principle and International Law: The Challenge of Implementation*, The Hague: Kluwer Law International, 1996

Kiss, A., 'The Right to the Conservation of the Environment', in Picolotti, R. and Taillant, D. (eds), *Linking Human Rights and Environment*, Tuscon, Arizona: University of Arizona Press, 2003

Kiss, A. and Shelton, D., *Manual of European Environmental Law*, Cambridge: Grotius, 1993

Knodt, E., 'The Habermas/Luhmann Controversy Revisited', *New German Critique* 61, Winter Issue: 77–100, 1994

Knorr-Cetina, K., *Epistemic Culture: How the Sciences Make Knowledge*, Cambridge, MA: Harvard University Press, 1999

Knox, P. and Taylor, P. (eds), *World Cities in a World-System*. Cambridge: Cambridge University Press, 1995

Kockelmans, J., *Edmund Husserl's Phenomenology*, West Lafayette, Indiana: Purdue University Press, 1994

Koffman, E. and Youngs, G. (eds), *Globalization: Theory and Practice*, London: Pinter, 1996

Koppen, I., 'Ecological Covenants: Regulatory Informality in Dutch Waste Reduction Policy', in Teubner, G. *et al.* (eds), *Environmental Law and Ecological Responsibility: The Concept and Practice of Ecological Self-Organization*, Chichester: John Wiley & Sons, 1994

Krall, F. R., *Ecotone: Wayfaring on the Margins*, New York: State University of New York Press, 1994

Krämer, L., 'The Citizen in the Environment: Access to Justice', 8 *Environmental Liability* 5: 127–41, 2000

Krohn, W., Kuppers, G. and Nowotny, H. (eds), *Self-organization: Portrait of a Scientific Evolution*, Dordrecht: Kluwer Acedemic, 1990

Ksentini, 'UN Commission on Human Rights, Final Report of the Special Rapporteur', UN Doc. E/CN 4/Sub.2/1994/9, 1994

La Cour, A., 'The Concept of Environment in Systems Theory', 13 *Cybernetics and Human Knowing* 1, 2006

Lacey, N., *Unspeakable Subjects: Feminist Essays in Legal and Social Theory*, Oxford: Hart Publishing, 1998

Ladeur, K-H., 'Coping with Uncertainty: Ecological Risks and the Proceduralization of Environmental Law', in Teubner, G. *et al.* (eds), *Environmental Law and Ecological Responsibility: The Concept and Practice of Ecological Self-Organization*, Chichester: John Wiley & Sons, 1994

Ladeur, K-H., *Social Risks, Welfare Rights and the Paradigm of Proceduralisation*, 95/2 EUI Working Paper, Firenze: European University Institute, 1995a

Ladeur, K-H., *Post-Modern Constitutional Theory: A Prospect for the Self-Organizing Society*, 95/6 EUI Working Paper, Firenze: European University Institute, 1995b

Ladeur, K-H., *The Theory of Autopoiesis as an Approach to a Better Understanding of Postmodern Law: From the Hierarchy of Norms to the Heterarchy of Changing Patterns of Legal Inter-Relationships*, 99/3 EUI Working Paper, Firenze: European University Institute, 1999

Ladeur, K-H. and Prelle, R., 'Environmental Assessment and Judicial Approaches to Procedural Errors – A European and Comparative Analysis', 13 *Journal of Environmental Law* 2: 186–98, 2001

Langsdorf, L., Watson, S. and Bower, M. (eds), *Phenomenology, Interpretation, and Community*, New York: State University of New York Press, 1996

Lasch, C., *The True and Only Heaven*, New York: Norton, 1991

Lash, S., Szerszynski, B., and Wynne, B. (eds), *Risk, Environment & Modernity: Towards a New Ecology*, London: Sage, 1996

Latour, B., *We Have Never Been Modern*, trans. C. Porter, Cambridge, MA: Harvard University Press, 1993

Latour, B., *Politics of Nature*, Cambridge, MA: Harvard University Press, 2004a

Latour, B., 'How to Talk about the Body: The Normative Dimension of Science Studies', 10 *Body & Society* 2–3: 205–29, 2004b

Lau, M., 'Islam and Judicial Activism: Public Interest Litigation and Environmental Protection in the Islamic Republic of Pakistan', in Boyle, A. and Anderson, M. (eds), *Human Rights Approaches to Environmental Protection*, Oxford: Clarendon Press, 1996

Laurence, D., *Waste Regulation Law*, London: Butterworths, 1999

Lechte, J., '(Not) Belonging in Postmodern Space', in Watson, S. and Gibson, K. (eds), *Postmodern Cities and Spaces*, Oxford: Blackwell, 1995

Lees, L. (ed.), *The Emancipatory City*, London: Sage, 2004

Lefebvre, H., *The Production of Space*, trans. D. Nicholson-Smith, Oxford: Blackwell, 1991

Lefebvre, H., *Writings on Cities*, ed. and trans. E. Koffman and G. Youngs, *Globalization: Theory and Practice*, London: Pinter, 1996

Lempert, R., 'The Autonomy of Law: Two Visions Compared', in Teubner, G. (ed.), *Autopoietic Law: A New Approach to Law and Society*, Berlin: Walter de Gruyter, 1988

Leopold, A., *Round River*, Oxford: Oxford University Press, 1953

Lévinas, E., *Totality and Infinity*, trans. A. Lingis, The Hague: Martinus Nijhoff, 1979

Lévinas, E., *Otherwise than Being or Beyond Essence*, trans. A. Lingis, The Hague: Martinus Nijhoff, 1981

Lévinas, E., *Hors Sujet*, Paris: LGF, 1997

Lewis, C., 'Women, Body, Space: Rio Carnival and the Politics of Performance', 3 *Gender Place and Culture* 1: 23–42, 1996

Locke, C., 'Digital Memory and the Problem of Forgetting', in Radstone, S. (ed.), *Memory and Methodology*, Oxford: Berg, 2000

Lopez, L., 'Between Hope and Fear: the Psychology of Risk', 20 *Advances in Experimental Psychology*, 1987

Lovelock, J., 'Taking Care', in O'Riordan, T. and Cameron, J. (eds), *Interpreting the Precautionary Principle*, London: Earthscan, 1994

Low, M., *Representation Unbound: Globalization and Democracy*, in Cox, K. (ed.), *Spaces of Globalization: Reasserting the Power of the Local*, New York and London: The Guildford Press, 1997

Lowenthal, L., 'The Utopian Motif is Suspended', in Jay, M. (ed.), *An Unmastered Past: the Autobiographical Reflections of Leo Lowenthal*, Berkeley: University of California Press, 1987

Luhmann, N., *The Differentiation of Society*, trans. S. Holmes and C. Larmore, NY: Columbia University Press, 1982a

Luhmann, N., 'Autopoiesis, Handlung, und Kommunikative Verständigung', 11 *Zeitschrift für Soziologie* 366, 1982b

Luhmann, N., 'World Society as a Social System' in Geyer, F. and van der Zouwen, J. (eds), *Dependence and Equality: A Systems Approach to the Problems of Mexico and Other Developing Countries*, Oxford: Oxford University Press, 1982c

Luhmann, N., *A Sociological Theory of Law*, trans. E. King and M. Albrow, Boston: Routledge & Kegan Paul, 1985a

Luhmann, N., 'The Self-Production of the Law and its Limits', in Teubner, G. (ed.), *Dilemmas of Law in the Welfare State*, Berlin: Walter De Gruyter, 1985b

Luhmann, N., 'The Autopoiesis of Social Systems', in Geyer, F. and van der Zouwen, J. (eds), *Sociocybernetic Paradoxes: Observation, Control and Evolution of Self-Steering Systems*, Beverly Hills: Sage, 1986a

Luhmann, N., 'The Coding of the Legal System', in Teubner, G. and Febbrajo, A. (eds), *State, Law, Economy as Autopoietic Systems*, Milano: Giuffré, 1986b

Luhmann, N., 'Some Problems with Reflexive Law', in Teubner, G. and Febbrajo, A. (eds), *State, Law and Economy as Autopoietic Systems*, Milan: Guiffré, 1986c

Luhmann, N., 'Closure and Openness: On Reality in the World of Law', in Teubner, G. (ed.) *Autopoietic Law: A New Approach to Law and Society*, Berlin: de Gruyter, 1988a

Luhmann, N., 'The Unity of the Legal System' in Teubner, G. (ed.), *Autopoietic Law: A New Approach to Law and Society*, Berlin: de Gruyter, 1988b

Luhmann, N., 'Law as a Social System', 83 *Northwestern University Law Review*: 136–50, 1989a

Luhmann, N., *Ecological Communication*, trans. J. Bednarz, Jr., Cambridge: Polity Press, 1989b

Luhmann, N., *Political Theory in the Welfare State*, trans J. Bednarz, Jr., Berlin: Walter de Gruyter, 1990a

Luhmann, N., 'The Cognitive Program of Constructivism and a Reality that Remains Unknown', in Krohn, W. *et al.* (eds), *Self-organization: Portrait of a Scientific Evolution*, Dordrecht: Kluwer Acedemic, 1990b

Luhmann, N., *Essays on Self-Reference*, New York: Columbia University Press, 1990c

Luhmann, N., 'Sthenography', 7 *Stanford Literature Review*: 133–7, 1990d

Luhmann, N., *Die Wissenschaft der Gesellschaft*, Vol. II, Frankfurt am Main: Suhrkamp, 1990e

Luhmann, N., 'Paradigm Lost: On the Ethical Reflections of Morality', *Thesis Eleven* 29: 82–94, 1991

Luhmann, N., 'Closure and Structural Coupling', 13 *Cardozo Law Review*: 1419–42, 1992a

Luhmann, N., 'Zur Einführung' in Neves, M. (ed.), *Verfassung und Positivität des Rechts in der peripheren Moderne. Eine theoretische Betrachtung und Interpretation des Falles Brasilien*, Berlin: Duncker & Humblot, 1992b

Luhmann, N., *Risk: A Sociological Theory*, trans. R. Barrett, New York: Aldine de Gruyter, 1993a

Luhmann, N., *Das Recht der Gesellschaft*, Frankfurt am Main: Suhrkamp, 1993b

Luhmann, N., 'Observing Re-entries', 16 *Graduate Faculty Philosophy Journal*: 485–98, 1993c

Luhmann, N., 'Deconstruction as Second-Order Observing', *New Literary History* 24: 763–82, 1993d

Luhmann, N., 'Speaking and Silence', trans. K. Behnke, 61 *New German Critique*, Winter Issue: 25–38, 1994a

Luhmann, N., 'The Modernity of Science', trans. K. Behnke, 61 *New German Critique*, Winter Issue: 9–24, 1994b

Luhmann, N., 'How Can the Mind Participate in Communication?', in Gumbrecht, H. U. and Pfeiffer, K. L. (eds), *Materialities of Communication*, trans. W. Whobrey, Stanford, California: Stanford University Press, 1994c

Luhmann, N., *Social Systems*, trans. J. Bednarz, Jr., Stanford, California: Stanford University Press, 1995a

Luhmann, N., 'Legal Argumentation: An Analysis of its Form', trans. I. Fraser, ed. W. T. Murphy and G. Teubner, 58 *Modern Law Review* 3: 285–98, 1995b

Luhmann, N., Hayles, K., Rasch, W., Knodt, E. and Wolfe, C., 'Theory of a Different Order: A Conversation with Katherine Hayles and Niklas Luhmann', *Cultural Critique* 31: 7–36, 1995c

Luhmann, N., 'Inklusion und Exklusion' in Luhmann, N. (ed.), *Soziologische Aufklärung. Vol 6: Die Soziologie und der Mensch*, Opladen: Westdeutscher, 1995d

Luhmann, N., 'Why Does Society Describe Itself as Postmodern?', *Cultural Critique* 30: 180–97, 1995e

Luhmann, N., *Gesellschaftstruktur und Semantik. Studien Zur Wissenssoziologie der modernen Gesellschaft*, Vol. 3, Frankfurt am Main: Suhrkamp, 1995f

Luhmann, N., *Protest: Systemtheorie und Soziale Bewegungen*, Frankfurt: Suhrkamp, 1996a

Luhmann, N., 'Complexity, Structural Contingencies and Value Conflicts', in Heelas, P. *et al.* (eds), *Detraditionalization: Critical Reflections on Authority and Identity*, Blackwell: Oxford, 1996b

Luhmann, N., *Die Gesellschaft der Gesellschaft*. Frankfurt am Main: Suhrkamp, 1997a

Luhmann, N., 'Limits of Steering', trans. and ed. M. Hohlweck and J. Paterson, 14 *Theory, Culture & Society* 1: 41–57, 1997b

Luhmann, N., *Observations on Modernity*, trans. W. Whobney, Stanford, California: Stanford University Press, 1998a

Luhmann, N., *Love as Passion: The Codification of Intimacy*, trans. J. Gaines and D. Jones, Stanford, California: Stanford University Press, 1998b

Luhmann, N., *Grundrechte als Institution*, Berlin: Dunker und Humblot, 1999

Luhmann, N., *Art as a Social System*, trans. E. Knodt, Stanford, California: Stanford University Press, 2000a

Luhmann, N., *Die Religion der Gesellschaft*, Frankfurt: Suhrkamp, 2000b

Luhmann, N., 'Notes on the Project "Poetry and Social Theory" ', 18 *Theory, Culture & Society* 1: 15–28, 2001

Luhmann, N., *Theories of Distinction: Redescribing the Descriptions of Modernity*, ed. and introduced by W. Rasch, Stanford, California: Stanford University Press, 2002a

Luhmann, N., 'Identity-What or How?' in Luhmann, N., *Theories of Distinction: Redescribing the Descriptions of Modernity*, ed. and introduced by W. Rasch, Stanford, California: Stanford University Press, 2002b

Luhmann, N., 'The Modern Sciences and Phenomenology' in Luhmann, N., *Theories of Distinction: Redescribing the Descriptions of Modernity*, ed. and introduced by W. Rasch, Stanford, California: Stanford University Press, 2002c

Luhmann, N. 'The Cognitive Program of Constructivism and the Reality that Remains Unknown', in Luhmann, N., *Theories of Distinction: Redescribing the Descriptions of Modernity*, ed. and introduced by W. Rasch, Stanford, California: Stanford University Press, 2002d

Luhmann, N., 'The Paradox of Observing Systems', in Luhmann, N., *Theories of Distinction: Redescribing the Descriptions of Modernity*, ed. and introduced by W. Rasch, Stanford, California: Stanford University Press, 2002e

Luhmann, N., 'Deconstruction as Second-Order Observation', in Luhmann, N., *Theories of Distinction: Redescribing the Descriptions of Modernity*, ed. and introduced by W. Rasch, Stanford, California: Stanford University Press, 2002f

Luhmann, N., *Law as a Social System*, trans. K Ziegert, eds. F. Kastner, R. Nobles, D. Schiff and R. Ziegert, Oxford: Oxford University Press, 2004

Luke, T., 'Identity, Meaning and Globalization: Detraditionalization in Postmodern Space-Time Compression', in Heelas, P. *et al.* (eds), *Detraditionalization: Critical Reflections on Authority and Identity*, Blackwell: Oxford, 1996

Lyotard, J-F., *The Postmodern Condition: a Report on Knowledge*, trans. G. Bennington and B. Massumi, Manchester: Manchester University Press, 1984

Lyotard, J-F., *The Differend: Phrases in Dispute*, trans. G. van den Abbeele, Minneapolis: University of Minnesota Press, 1988

Lyotard, J-F., *Political Writings*, trans. B. Readings and K. Geiman, Minneapolis: University of Minnesota Press, 1993

Lyotard, J-F., *Lessons on the Analytic of the Sublime*, trans. E. Rottenberg, Stanford: Stanford University Press, 1994

MacGarvin, M., 'Precaution, Science and the Sin of Hubris', in O' Riordan, T. and Cameron, J. (eds), *Interpreting the Precautionary Principle*, London: Earthscan, 1994

MacIntyre, A., *After Virtue: A Study in Moral Theory*, London: Duckworth, 1985

McAuslan, P., *Urban Land and Shelter for the Poor*, Nottingham: Earthscan, 1985

McAuslan, P., *Bringing the Law Back In: Essays in Land, Law and Development*, Ashgate: Aldershot, 2003

McClymonds, J., 'The Human Right to a Healthy Environment: an International Perspective', 37 *New York School Law Review*: 583–633, 1992

McIntyre, O., 'The All-Consuming Definition of "Waste" and the End of the "Contaminated Land" Debate?', 17 *Journal of Environmental Law* 1: 109–27, 2005

McKibben, B., *The End of Nature*, London: Penguin, 1990

Macann, C., *Presence and Coincidence: The Transformation of Transcendental into Ontological Phenomenology*, Phaenomenologica 119, Dordrecht: Kluwer, 1991

Macann, C., *Four Phenomenological Philosophers*, London: Routledge, 1993

Madon, J.P., 'The Law, The Heart: Blanchot and the Question of Community', in Pepper, T. (ed.), *The Place of Maurice Blanchot, 93 Yale French Studies*: 60–6, 1998

Maguire, J., 'The Tears Inside the Stone: Reflections on the Ecology of Fear', in Lash, S. *et al.* (eds), *Risk, Environment & Modernity: Towards a New Ecology*, London: Sage, 1996

Manderson, D., 'Interstices: New Work on Legal Spaces', 9 *Law Text Culture*: 1–11, 2005

Manuel, F. E. and Manuel, F. P., *French Utopias: An Anthology of Ideal Societies*, New York: Free Press, 1966

Manuel, F. E. and Manuel, F. P., *Utopian Thought in the Western World*, Oxford: Blackwell, 1979

Markandya, A. and Richardson, J. (eds), *The Earthscan Reader in Environmental Economics*, London: Earthscan, 1992

Massey, D., 'Politics and Space/Time', 196 *New Left Review*: 65–84, 1992

Massey, D., *Space, Place and Gender*, Cambridge: Polity Press, 1994

Massey, D., 'The Conceptualization of Place', in Massey, D. and Jess, P. (eds), *A Place in the World? Places, Cultures and Globalisation*, Oxford: Open University Press, 1995

Massey, D., 'Cities in the World' in Massey, D. *et al.* (eds), *City Worlds*, London: Routledge, 1999

Massey, D., Allen, J. and Pile, S. (eds), *City Worlds*, London: Routledge, 1999

Massey, D. and Jess, P. (eds), *A Place in the World? Places, Cultures and Globalisation*, Oxford: Open University Press, 1995

Maturana, H., 'The Organization of the Living: A Theory of the Living Organization', 51 *International Journal of Human and Computer Studies* 2: 149–68, 1999

Maturana, H. and Varela, F., *Autopoiesis and Cognition: the Realization of the Living*, Dordecht, Holland: Reidel Publishing, 1972

Maturana, H. and Varela, F., *The Tree of Knowledge: the Biological Roots of Human Understanding*, Boston: Shambala, 1992

May, J. and Thrift, N. (eds), *Timespace: Geographies of Temporality*, London: Routledge, 2001

Mazlish, B., 'The *Flâneur*: from Spectator to Representation', in Tester, K. (ed.), *The Flâneur*, London: Routledge, 1994

Merchant, C., *The Death of Nature: Women Ecology and the Scientific Revolution*, London: Wildwood House, 1980

Merleau-Ponty, M., *The Phenomenology of Perception*, London: Routledge & Kegan Paul, 1962

Merrifield, A. and Swyngedouw, E. (eds), *The Urbanization of Justice*, London: Lawrence and Wishart, 1996

Merrills, J. G., 'Environmental Protection and Human Rights: Conceptual Aspects', in Boyle, A. and Anderson, M. (eds), *Human Rights Approaches to Environmental Protection*, Oxford: Clarendon Press, 1996

Meyer, A., 'International Environmental Law and Human Rights: Towards the Explicit Recognition of Traditional Knowledge', 10 *RECIEL* 1: 37–46, 2001

Mitchell, D., 'The Annihilation of Space by Law: The Roots and Implications of Anti-Homeless Laws in the US', 29 *Antipode* 3: 303–35, 1997

Mitchell, W., 'Building the Bitsphere, or the Kneebone's Connected to the Bahn', *ID Magazine*: 13–18, November 1994

Mitchell, W., *City of Bits: Space, Place and the Infobahn*, Cambridge, MA: MIT Press, 1995

Mohanty, J. H., 'The Development of Husserl's Thought', in Smith, B. and Woodruff Smith, D. (eds), *The Cambridge Companion to Husserl*, Cambridge: Cambridge University Press, 1995

Moltke, K., von, 'The Relationship Between Policy, Science, Technology, Economics and Law in the Implementation of the Precautionary Principle', in Freestone, D. and Hey, E. (eds), *The Precautionary Principle and International Law: The Challenge of Implementation*, The Hague: Kluwer Law International, 1996

Morin, E., *La Méthode: La Nature de la Nature*, Paris: Seuil, 1977

Morin, E., *La Méthode II: La Connaissance de la Connaissance*, Paris: Seuil, 1986

Morin E., 'From the Concept of System to the Paradigm of Complexity', 15 *Journal of Social and Evolutionary Systems* 4: 371–85, 1992

Motha, S., 'The Failure of Postcolonial Sovereignty in Australia', 22 *Australian Feminist Law Journal*: 107–25, 2005

Motha, S. and T. Zartaloudis, 'Law, Ethics and the Utopian End of Human Rights', 12 *Social and Legal Studies* 2: 243–68, 2003

Mumford, L., *The Story of Utopias*, London: G. Harrap, 1923

Mumford, L., *The City in History*, London: Penguin, 1961

Munch, R., 'Autopoiesis by Definition', 13 *Cardozo Law Review* 5: 1463–71, 1992

Murphy, W. T., 'Modern Times: Niklas Luhmann on Law, Politics and Social Theory', 47 *The Modern Law Review* 5: 603–21, 1984

Murphy, W. T., 'Some Issues in the Relationship between Law and Autopoiesis', 5 *Law and Critique* 2: 241–64, 1994a

Murphy, W.T., 'As If: Camera Juridica', in Douzinas, C. *et al.* (eds), *Politics, Postmodernity and Critical Legal Studies: The Legality of the Contingent*, London: Routledge, 1994b

Murphy, W.T., *The Oldest Social Science? Configurations of Law and Modernity*, Oxford: Clarendon, 1997

Murphy, W.T., 'Modernising Justice Inside "UK PLC": Mimesis, De-differentiation and Colonisation', in Pribáñ, J. and Nelken, D. (eds), *Law's New Boundaries: The Consequences of Legal Autopoiesis*, Aldershot: Dartmouth Ashgate, 2001

Naess, A., *Ecology, Community, and Lifestyle: Outline of an Ecosophy*, trans. D. Rothenberg, Cambridge: Cambridge University Press, 1989

Nancy, J-L., *The Inoperative Community*, trans. P. Connor, L. Garbus, M. Holland, S. Sawhney, ed. P. Connor, Minneapolis: Minnesota University Press, 1991

Nancy, J-L., *Being Singular Plural*, trans. R. Richardson and A. O'Byrne, Stanford: Stanford University Press, 2000

Nankhnikian, G., 'Introduction', in Husserl, E. *The Idea of Phenomenology*, trans. W. Alston and G. Nakhnikian, The Hague: Martinus Nijhoff, 1964

Napier, C., *Waste Management: Legal Requirements and Good Practice for Producers of Waste*, ed. A. Roney, Oxford: Chandos, 1998

Nelken, D. (ed.), *Law as Communication*, Aldershot: Dartmouth, 1996

Nelson, L., 'Bodies (and Spaces) Do Matter: The Limits of Performativity', *Gender, Place and Culture* 6: 331–53, 1999

Neves, M., 'From the Autopoiesis to the Allopoiesis of Law', 28 *Journal of Law and Society* 2: 242–64, 2001

Nicholson, L. (ed.), *Feminism/Postmodernism*, London: Routledge, 1990

Nivola, P., *Laws of the Landscape: How Policies Shape Cities in Europe and America*, Washington DC: Brookings Institution Press, 1999

Nollkaemper, A., ' "What You Risk Is What You Value", and Other Dilemmas Encountered in the Legal Assaults on Risks', in Freestone, D. and Hey, E. (eds), *The Precautionary Principle and International Law: The Challenge of Implementation*, The Hague: Kluwer Law International, 1996

Nora, P. (ed.), *Les Lieux de Mémoire*, Paris: Gallimard, 1984–1993

Norrie, A. (ed.), *Closure or Critique: New Directions in Legal Theory*, Edinburgh: Edinburgh University Press, 1993

Norrie, A., *Law and the Beautiful Soul*, London: Glasshouse Press, 2005

O' Riordan, T. and Cameron, J. (eds), *Interpreting the Precautionary Principle*, London: Earthscan, 1994

Odum, H., *Environment, Power, and Society*, New York: John Wiley and Sons, 1971

Odum, H. and Odum, E., *Energy Basis for Man and Nature*, New York: McGraw-Hill, 1981

OECD, *Environmental Policies for Cities in the 1990s*, Paris: OECD, 1990

Orts, E., 'Reflexive Environmental Law', 89 *Northwestern University Law Review* 4: 1227–79, 1994

Orts, E., 'Autopoiesis and the Natural Environment', in Pribáñ, J. and Nelken, D. (eds), *Law's New Boundaries: The Consequences of Legal Autopoiesis*, Aldershot: Dartmouth Ashgate, 2001

Pachoud, B., 'The Teleological Dimension of Perceptual and Motor Intentionality', in Petitot, J. *et al.* (eds), *Naturalizing Phenomenology: Issues in Contemporary Phenomenology and Cognitive Science*, Stanford, CA: Stanford University Press, 1999

Pallemaerts, M., 'International Law and Sustainable Development: Any Progress in Johannesburg?', 12 *RECIEL* 1: 1–11, 2003

Panjabi, R. K., 'Idealism and Self-Interest in International Law: The Rio Dilemma', *Californian Western International Law Journal* 23: 189–97, 1992

Passet, R., *L'Economie et le Vivant*, Paris: Payot, 1979

Paterson, J., 'Who is Zenon Bankowski Talking To? The Person in the Sight of Autopoiesis', in D. Nelken (ed.), *Law as Communication*, Aldershot: Dartmouth, 1996

Paterson, J., 'Trans-Science, Trans-Law and Proceduralization', 12 *Social & Legal Studies* 4: 525–45, 2003

Paterson, J., 'Reflecting on Reflexive Law', in King, M. and Thornhill, C. (eds) *Luhmann on Politics and Law: Critical Appraisals and Applications*, Oxford: Hart, 2006

Paterson, J. and Teubner, G., 'Changing Maps: Empirical Legal Autopoiesis', 7 *Social & Legal Studies* 4: 451–86, 1998

Patten, K., 'Teaching "Discovering Silence" ', in Jarowski, A. (ed.), *Silence: Interdisciplinary Perspectives*, Berlin: Mouton de Gruyter, 1997

Perez, O., *Ecological Sensitivity and Global Legal Pluralism: Rethinking the Trade and Environment Conflict*, Oxford: Hart, 2004

Perez, O. and Teubner, G. (eds), *Paradoxes and Inconsistencies in Law*, Oxford: Hart, 2006

Perrin, C., 'Breath from Nowhere: The Silent "Foundation" of Human Rights', 13 *Social and Legal Studies* 1: 133–51, 2004

Petit, J.L., 'Constitution by Movement: Husserl in the Light of Recent Neuro-biological Findings', in Petitot, J. *et al.* (eds), *Naturalizing Phenomenology: Issues in Contemporary Phenomenology and Cognitive Science*, Stanford, CA: Stanford University Press, 1999

Petitot, J., Varela, F., Pachoud, B. and Roy. J-M. (eds), *Naturalizing Phenomenology: Issues in Contemporary Phenomenology and Cognitive Science*, Stanford, CA: Stanford University Press, 1999

Philippopoulos-Mihalopoulos, A., 'Mapping Utopias: A Voyage to Placelessness', 2 *Law and Critique* 12: 135–57, 2001

Philippopoulos-Mihalopoulos, A., 'Suspension of Suspension: Settling for the Improbable', 15 *Law and Literature* 3: 345–70, 2003

Philippopoulos-Mihalopoulos, A., 'Between Law and Justice: A Connection of No Connection between Luhmann and Derrida', in Himma, K. E. (ed.), *Law, Morality, and Legal Positivism* (ARSP Beihefte), Stuttgart: Franz Steiner Verlag, 2004a

Philippopoulos-Mihalopoulos, A., 'Caspian Catachreses: Environmental Transplant-ing in a Space of Flows', in Paterson, J., Bantekas, I. and Suleimenov, S. (eds), *Kazakhstan Oil and Gas Law: National and International Perspectives*, The Hague: Kluwer, 2004b

Philippopoulos-Mihalopoulos, A., 'Beauty and the Beast: Art and Law in the Hall of Mirrors', 2 *Entertainment Law* 3: 1–34, 2004c

Philippopoulos-Mihalopoulos, A., 'Between Light and Darkness: Earthsea and the Name of Utopia', 8 *Contemporary Justice Review* 1: 45–57, 2005a

Philippopoulos-Mihalopoulos, A., 'Dealing (with) Paradoxes: On Law, Justice and Cheating' in King, M. and Thornhill, C. (eds), *Luhmann on Law and Politics: Critical Appraisals and Applications*, Oxford: Hart, 2005b

Philippopoulos-Mihalopoulos, A., '*Before*: Gender, Identity, Human Rights', 14 *Feminist Legal Studies* 3: 271–91, 2006

Philippopoulos-Mihalopoulos, A. (ed.), *Law and the City*, London: Routledge-Cavendish, 2007

Philipse, H., 'Transcendental Idealism', in Smith, B. and Woodruff Smith, D. (eds), *The Cambridge Companion to Husserl*, Cambridge: Cambridge University Press, 1995

Pickering, J., 'Designs on the City: Urban Experience in the Age of Electronic Reproduction', in Downey, J. and McGuigan, J., *Technocities*, London: Sage, 1999

Picolotti, R. and Taillant, D. (eds), *Linking Human Rights and Environment*, Tuscon, Arizona: University of Arizona Press, 2003

Pile, S. and Thrift, N. (eds), *Mapping the Subject: Geographies of Cultural Transform-ation*, London: Routledge, 1995

Pile, S., Brook, C. and Mooney, G. (eds), *Unruly Cities? Order/Disorder*, London: Routledge, 1999

Plant, S., *Zeros and Ones: Digital Women and the New Technoculture*, London: Routledge, 1997

Plumwood, V., *Feminism and the Mastery of Nature*, London: Routledge, 1993
Plumwood, V., 'Inequality, Ecojustice and Ecological Rationality', in Dryzek, J. and Schlosberg, D. (eds), *Debating the Earth: the Environmental Politics Reader*, Oxford: Oxford University Press, 1998
Pocklington, D., *The Law of Waste Management*, Crayford: Shaw and Sons, 1997
Podgorecki, A., Whelan, C. and Khosla, D., *Legal Systems & Social Systems*, London: Croom Helm, 1985
Potter, R. and Lloyd-Evans, S., *The City in the Developing World*, Harlow: Longman, 1998
Pratt, G., 'Grids of Difference: Place and Identity Formation', in Fincher, R. and Jacobs, J. (eds), *Cities of Difference*, London: The Guildford Press, 1998
Pribáñ, J., 'Legitimation between the Noise of Politics and the Order of Law', in Pribáñ, J. and Nelken, D. (eds), *Law's New Boundaries: The Consequences of Legal Autopoiesis*, Aldershot: Dartmouth Ashgate, 2001
Prigogine, I. and Stengers, I., *Order out of Chaos*, New York: Bantham, 1984
Pugh, A., *Silent Reading: An Introduction to its Study and Teaching*, London: Heinemann, 1978
Raban, J., *Soft City*, London: Hamish Hamilton, 1974
Radstone, S., 'Screening Trauma: *Forrest Gump*, Film and Memory' in Radstone, S. (ed.), *Memory and Methodology*, Oxford: Berg, 2000
Rasch, W., *Niklas Luhmann's Modernity: The Paradoxes of Differentiation*, Stanford, California: Stanford University Press, 2000
Rasch, W., *Sovereignty and its Discontents: On the Primacy of Conflict and the Structure of the Political*, London: Birkbeck Law Press, 2004
Rasch, W. and Wolfe, C. (eds), *Observing Complexity: Systems Theory and Post-modernism*, Minneapolis, MN: University of Minnesota Press, 2000
Rawls, J., *A Theory of Justice*, Cambridge, MA: Harvard University Press, 1971
Redclift, M., *Sustainable Development: Exploring the Contradictions*, London: Routledge, 1991
Richardson, B., and Wood, S. (eds), *Reader on Environmental Law for Sustainability*, Oxford: Hart, 2006
Rilke, R. M., *Selected Poetry and Prose*, ed. and trans. S. Mitchell, New York: The Modern Library, 1995
Roberts, D. (ed.), *Reconstructing Theory: Gadamer, Habermas, Luhmann*, Victoria: Melbourne University Press, 1995
Robinson, D. and Dunkley, J. (eds), *Public Interest Perspectives in Environmental Law*, London: Wiley Chancery, 1995
Rodaway, P., *Sensuous Geographies: Body, Sense and Place*, London: Routledge, 1994
Rodenbach, G., *Bruges-la-Morte*, trans. M. Mitchell and W. Stone, Sawtry, Cambs: Dedalus, 2005
Rodriguez, D. and Torres, J., 'Autopoiesis, la Unidad de una Diferencia: Luhmann y Maturana', 5 *Sociologias* 9: 106–40, 2003
Rogowski, R., 'The Paradox of Law and Violence: Modern and Postmodern Readings of Benjamin's "Critique of Violence" ', 18 *New Comparison*, 131–51, 1994
Rolston III, H., *Environmental Ethics: Duties to and Values in the Natural World*, Philadelphia: Temple University Press, 1988
Rose, G., *Feminism and Geography: The Limits of Geographical Knowledge*, Minneapolis: University of Minnesota Press, 1993

Rose, G., *Mourning Becomes the Law: Philosophy and Representation*, Cambridge: Cambridge University Press, 1996

Rose, G., 'Performing Space' in Massey, D., Allen, J. and Sarre, P. (eds), *Human Geography Today*, Cambridge: Polity, 1999

Rottleuthner, H., 'Biological Metaphors in Legal Thought', in Teubner, G. (ed.), *Autopoietic Law: A New Approach to Law and Society*, Berlin: Walter de Gruyter, 1988

Rottleuthner, H., 'A Purified Theory of Law: Niklas Luhmann on the Autonomy of the Legal System', 23 *Law and Society Review*: 779–97, 1989

Ruddick, S., 'Domesticating Monsters: Cartographies of Difference and the Emancipatory City', in Lees, L. (ed.), *The Emancipatory City*, London: Sage, 2004

Sagoff, M., 'On Preserving the Natural Environment', 84 *Yale Law Review*: 205–67, 1974

Sagoff, M., *The Economy of the Earth: Philosophy, Law and Economics*, Cambridge University Press: Cambridge, 1988

de Saint-Exupéry, A., *Le Petit Prince*, Paris: Gallimard, 1946

Sand, I.-J., 'The Legal Regulation of the Environment and New Technologies – In View of Changing Relations between Law, Politics and Science: The Case of Applied Genetic Technology', 22 *Zeitschrift für Rechtssoziologie* 2: 1–38, 2001

Sand, I.-J., 'The Regulation of Vital Risks, Uncertainties and Scientific Controversies', in Damgaard, C., Henrichsen, C. and Petersen, H. (eds), *Ret & Usikkerhed*, Copenhagen: Jurist-og Økonomforbundets Forlag, 2005

Sandel, M., *Liberalism and the Limits of Justice*, Cambridge: Cambridge University Press, 1982

Sandilands, C., 'The Good-Natured Feminist: Ecofeminism and Democracy', in Keil, R. *et al.* (eds), *Political Ecology: Global and Local*, London: Routledge, 1998

Sands, P., *Principles of International Environmental Law: Frameworks, Standards and Implementation*, Manchester, NY: Manchester University Press, 1994

Santayana, G., *Scepticism and Animal Faith: Introduction to a System of Philosophy*, New York: Dover Publications, 1955

Sarat, A. ' "... The Law is All Over": Power, Resistance and the Legal Consciousness of the Welfare Poor', 2 *Yale Journal of Law and Humanities* 2: 343–80, 1990

Sassen, S., *The Global City*, Princeton: Princeton University Press, 1991

Sassen, S., *Cities in a World Economy*, Thousand Oaks: Pine Forge Press, 1994

Schütz, A., 'Desiring Society: Autopoiesis and the Paradigm of Mastership', 5 *Law and Critique* 2: 149–64, 1994

Schütz, A., 'The Twilight of Global Polis: On Losing Paradigms, Environing Systems and Observing World Society', in Teubner, G. (ed.), *Global Law without a State*, Aldershot: Ashgate, 1996

Schütz, A., 'The Law With and Against Luhmann, Legendre, Agamben', 11 *Law and Critique* 2: 107–36, 2000

Scholte, J.A., 'Beyond the Buzzword: Towards a Critical Theory of Globalization', in Koffman, E. and Youngs, G. (eds), *Globalization: Theory and Practice*, London: Pinter, 1996

Sczuchewycz, B., 'Silence in Ritual Communication', in Jarowski, A. (ed.), *Silence: Interdisciplinary Perspectives*, Berlin: Mouton de Gruyter, 1997

Sennett, R., *The Fall of Public Man*, New York: Alfred A. Knopf, 1976

Sennett, R., *The Conscience of the Eye: The Design and Social Life of Cities*, New York: Alfred A. Knopf, 1990

Sennett, R., *Flesh and Stone: The Body and the City in Western Civilization*, London: Allen Lane, 1995

Sennett, R., 'Growth and Failure: The New Political Economy and Its Culture', in Featherstone, M. and Lash, S. (eds), *Spaces of Culture*, London: Sage, 1999

Serres, M., *Hermes: Literature, Science and Philosophy*, trans. and ed. J. Harari and D. Bell, Baltimore: Johns Hopkins University Press, 1982

Serres, M., *The Natural Contract*, trans. E. MacArthur and W. Paulson, The University of Michigan Press: Ann Arbor [*Le Contrat Naturel*, Paris: Bourin, 1990], 1995

Serres, M., *The Birth of Physics*, trans. J. Hawkes, Manchester: Clinamen Press, 2001

Shelton, D., 'Human Rights, Environmental Rights and the Right to the Environment', 28 *Stanford Journal of International Law*: 103–38, 1991

Shelton, D., 'What Happened in Rio to Human Rights?', *Yearbook of Environmental Law*: 75–93, London: Graham & Trotman, 1992

Shelton, D., 'The Impact of Scientific Uncertainty on Environmental Law and Policy in the United States', in Freestone, D. and Hey, E. (eds), *The Precautionary Principle and International Law: The Challenge of Implementation*, The Hague: Kluwer Law International, 1996

Shelton, D., *Commitment and Compliance: The Role of Non-Binding Norms in the International Legal System*, Oxford: Oxford University Press, 2000

Shelton, D., 'The Environmental Jurisprudence of International Human Rights Tribunals', in Picolotti, R. and Taillant, D. (eds), *Linking Human Rights and Environment*, Tuscon, Arizona: University of Arizona Press, 2003

Shields, R., 'A Guide to Urban Representation and What to Do About It: Alternative Traditions of Urban Theory', in King, A. (ed.), *Re-presenting the City: Ethnicity, Capital and Culture in the 21ˢᵗ Century Metropolis*, London: Macmillan, 1996

Shields, R., 'Flow as a New Paradigm', 1 *Space and Culture* 1: 1–7, 1997

Sibley, D., *Geographies of Exclusion*, London: Routledge, 1995

Singer, L., 'Recalling a Community at Loose Ends', in The Miami Theory Collective (ed.), *Community at Loose Ends*, Minneapolis: University of Minnesota Press, 1991

Sløk, C., 'Niklas Luhmann's Ambiguity Towards Religion', 2 *Soziale Systeme*, 2005

Smith, B. and Woodruff Smith, D. (eds), *The Cambridge Companion to Husserl*, Cambridge: Cambridge University Press, 1995

Smith, C., 'Autopoietic Law and the "Epistemic Trap": A Case Study of Adoption and Contact', 31 *Journal of Law and Society* 3: 318–44, 2004

Smith, M., *Ignorance and Uncertainty*, New York: Springer, 1989

Smith, T., 'Environmental Law: Old Ways and New Directions', 27 *Loyola of Los Angeles Law Review*, 1077–92, 1994

Soja, E., *Postmodern Geographies: The Reassertion of Space in Critical Social Theory*, London: Verso, 1989

Somsen, H., *Protecting the European Environment: Enforcing EC Environmental Law*, London: Blackstone, 1996

Soper, K., *What is Nature? Culture, Politics and the Non-Human*, Oxford: Blackwell, 1995

de Sousa Santos, B. (ed. of volume), State Transformation, Legal Pluralism and Community Justice, 1 *Social and Legal Studies* 2, 1992

de Sousa Santos, B., *Toward a New Common Sense: Law, Science and Politics in the Paradigmatic Transition*, New York: Routledge, 1995

Spaargaren, G., Mol, A. and Buttel, F. (eds), *Environment and Global Modernity*, London: Sage, 2000

Spencer Brown, G., *Laws of Form*, London: George Allen and Unwin, 1969

Spivak, G. C., 'Can the Subaltern Speak?', in Nelson, C. and Grossberg, L. (eds), *Marxism and the Interpretation of Culture*, London: Macmillan, 1988

Spivak, G. C., 'Who Claims Alterity?', in Kruger, B. and Mariani, P. (eds), *Remaking History*, Seattle: Bay Press, 1989

Sraffa, P., *Production of Commodities by Means of Commodities: Prelude to a Critique of Economic Theory*, Cambridge: Cambridge University Press, 1960

Stallworthy, M., *Sustainability, Land Use and Environment: A Legal Analysis*, London: Cavendish, 2002

Stanley, N., 'Public Concern: the Decision-Maker's Dilemma', *Journal of Planning and Environmental Law*: 919–34, 1998

Steeves, P., 'Constituting the Transcendent Community', in Langsdorf, L. *et al.* (eds), *Phenomenology, Interpretation, and Community*, New York: State University of New York Press, 1996

Steiner, G., *After Babel*, Oxford: Oxford University Press, 1975

Stensvaag, J.-M., 'The Not So Fine Print of Environmental Law', 27 *Loyola of Los Angeles Law Review*, April 1994: 1093–103, 1994

Stone, C., *Should Trees Have Standing? Toward Legal Rights for Natural Objects*, Los Altos: William Kaufmann, 1974

Stone, C., *Earth and Other Ethics*, New York: Harper and Row, 1987

Stone, C., *Should Trees Have Standing and Other Essays on Law, Morals, and the Environment*, Dobbs Ferry, NY: Oceana, 1996

Strathern, M., *Reproducing the Future: Anthropology, Kinship, and the New Reproductive Technologies*, London: Routledge, 1992

Sunstein, C. R., *Risk and Reason: Safety, Law and the Environment*, Cambridge: Cambridge University Press, 2002

Tarlock, D., 'The Non-Equilibrium Paradigm in Ecology and the Partial Unravelling of Environmental Law', 27 *Loyola of Los Angeles Law Review* 3: 1121–44, 1994

Tarlock, D., 'Is There a There in Environmental Law?', 19 *Journal of Land Use* 2: 213–54, 2004

Taylor, C., *Sources of the Self*, Cambridge: Cambridge University Press, 1989

Taylor, C., *The Ethics of Authenticity*, Cambridge, MA: Harvard University Press, 1991

Taylor, W., *The Geography of Law: Landscape, Identity and Regulation*, Oxford: Hart, 2006

Tellegen, E. and Wolsink, M., *Society and Its Environment: An Introduction*, Amsterdam: Gordon and Breach, 1998

Tester, K. (ed.), *The Flâneur*, London: Routledge, 1994

Teubner, G. (ed.), *Dilemmas of Law in the Welfare State*, Berlin: Walter De Gruyter, 1985

Teubner, G. (ed.), *Autopoietic Law: A New Approach to Law and Society*, Berlin: Walter de Gruyter, 1988

Teubner, G., *How the Law Thinks: Toward a Constructivist Epistemology of Law*, 89/393 EUI Working Paper, Firenze: European University Institute, 1989

Teubner, G., 'The Two Faces of Janus: Rethinking Legal Pluralism', 13 *Cardozo Law Review* 5: 1443–63, 1992

Teubner, G., *Law as an Autopoietic System*, trans. A. Bankowska and R. Adler, Oxford: Blackwell, 1993

Teubner, G. (ed.), *Global Law without a State*, Aldershot: Ashgate, 1996

Teubner, G., 'The King's Many Bodies: The Self-Deconstruction of Law's Hierarchy', 31 *Law and Society Review*: 763–88, 1997

Teubner, G., 'Economics of Gift – Positivity of Justice: The Mutual Paranoia of Jacques Derrida and Niklas Luhmann', 18 *Theory, Culture & Society* 1: 29–47, 2001a

Teubner, G., 'Alienating Justice: On the Surplus Value of the Twelfth Camel', in Pribáñ, J. and Nelken, D. (eds), *Law's New Boundaries: The Consequences of Legal Autopoiesis*, Aldershot: Dartmouth Ashgate, 2001b

Teubner, G. and Febbrajo, A. (eds), *State, Law, Economy as Autopoietic Systems*, Milano: Giuffré, 1986

Teubner, G., Farmer, L. and Murphy, D. (eds), *Environmental Law and Ecological Responsibility: The Concept and Practice of Ecological Self-Organization*, Chichester: John Wiley & Sons, 1994

Theunissen, M., *The Other: Studies in the Social Ontology of Husserl, Heidegger, Sartre, and Buber*, trans. C. Macann, Cambridge, MA: MIT Press, 1984

Thorme, M., 'Establishing Environment as Human Right', 19 *The Denver Journal of International Law and Policy* 2: 301–42, 1991

Thrift. N., *Spatial Formations*, London: Sage, 1996

Tönnies, F., *Community and Society*, trans. C. Loomis, New York: Harper, 1963

Toon, I., 'Finding a Place in the Street: CCTV Surveillance and Young People's Use of Urban Public Space', in Bell, D. and Haddour, A. (eds), *City Visions*, Harlow: Prentice Hall, 2000

Tournier, M., *Vendredi ou les Limbes du Pacific*, Paris: Gallimard (Folio), 1972

Tromans, S., 'Alternatives to Landfill – Can the Planning System Deliver?', *Journal of Planning and Environmental Law*: 257–64, March 2001

Tuan, Y-F., *Space and Place: The Perspective of Experience*, London: Edward Arnold, 1977

UNCED (United Nations Conference on Environment and Development), *Agenda 21*, Rio, 1992

Vago, S., *Law and Society*, 2nd Edition, New Jersey: Prentice Hall, 1988

Valverde, M., 'Taking "Land Use" Seriously: Toward an Ontology of Municipal Law', 9 *Law Text Culture*: 34–59, 2005

Valverde, M., 'Toronto: A "Multicultural" Urban Order', in Philippopoulos-Mihalopoulos, A. (ed.), *Law and the City*, London: Glasshouse Press, 2006

Van Zandt, D., 'The Breath of Life in the Law', 13 *Cardozo Law Review* 5: 1745–63, 1992

Vanderstraeten, R., 'Autopoiesis and Socialization: on Luhmann's Reconceptualization of Communication and Socialization', 51 *British Journal of Sociology* 3: 581–98, 2000

Verschraegen, G., 'Human Rights and Modern Society: A Sociological Analysis from the Perspective of Systems Theory', 29 *Journal of Law and Society* 2: 258–81, 2002

Villanueva Gardner, C., 'An Ecofeminist Perspective on the Urban Environment', in

Bennett, M. and Teague, D. (eds), *The Nature of Cities: Ecocriticism and Urban Environments*, Tuscon: University of Arizona Press, 1999

Virilio, P., 'The Third Interval', in Andermatt Conley, V., *Rethinking Technologies*, Minneapolis: University of Minnesota Press, 1993

Virilio, P., 'Un Monde Superexposé: Fin de l'Histoire, ou Fin de la Géographie?', *Le Monde Diplomatique*, August 1997

Vogel, L., *The Fragile 'We': Ethical Implications of Heidegger's Being and Time*, Evanston, Illinois: Northwestern University Press, 1994

Wagner, G., 'The End of Luhmann's System Theory', 27 *Philosophy of the Social Sciences*: 387–409, 1997

Wall, T. C., *Radical Passivity: Levinas, Blanchot, and Agamben*, New York: State University of New York Press, 1999

Warner, F., 'What if? Versus If It Ain't Broke, Don't Fix It', in O' Riordan, T. and Cameron, J. (eds), *Interpreting the Precautionary Principle*, London: Earthscan, 1994

Watson, S. and Gibson, K. (eds), *Postmodern Cities and Spaces*, Oxford: Blackwell, 1995

Weisberg, R., 'Autopoiesis and Positivism', 13 *Cardozo Law Review* 5: 1721–9, 1992

Weiss, E., *In Fairness to Future Generations*, Tokyo: United Nations University, 1989

Weiss, G., 'City Limits', 9 *City* 2: 215–24, 2005

Wenz, S., *Environmental Ethics Today*, Oxford: Oxford University Press, 2001

Westwood, S. and Williams, J. (eds), *Imagining Cities: Scripts, Signs, Memory*, London: Routledge, 1997

Wiener, J., 'Law and the New Ecology: Evolution, Categories and Consequences', 22 *Ecology Law Quarterly*: 325–57, 1995

Wilkinson, D., 'Using Environmental Ethics to Create Ecological Law', in Holder, J. and McGillivray, D. (eds), *Locality and Identity: Environmental Issues in Law and Society*, Aldershot: Ashgate, 1999

Withers, S., 'Silence and Communication in Art', in Jarowski, A. (ed.), *Silence: Interdisciplinary Perspectives*, Berlin: Mouton de Gruyter, 1997

Wolfe, A., 'Sociological Theory in the Absence of People: the Limits of Luhmann's Systems Theory', 13 *Cardozo Law Review* 5: 1729–45, 1992

Wolfe, C., *Critical Environments: Postmodern Theory and the Pragmatics of the 'Outside'*, Minneapolis: University of Minnesota Press, 1998

Wolfe, C., 'In Search of Posthumanist Theory: The Second-Order Cybernetics of Maturana and Varela', in Rasch, W. and Wolfe, C. (eds), *Observing Complexity: Systems Theory and Postmodernism*, Minneapolis, MN: University of Minnesota Press, 2000

Wolff, J., 'The Invisible *Flâneuse*, Women and the Literature of Modernity', in *Feminine Sentences: Essays on Women and Culture*, Berkeley: University of California Press, 1990

World Commission on Environment and Development (WCED), *Our Common Future*, Oxford: Oxford University Press, 1987

Wright, F. L., 'City of the Future', in Blowers, A. *et al.* (eds), *The Future of Cities*, London: Hutchinson, 1974

Wynne, B., 'Uncertainty and Environmental Learning: Reconceiving Science and Policy in the Preventative Paradigm', 2 *Global Environmental Change*: 111–27, 1992

Wynne, B., 'May the Sheep Safely Graze? A Reflexive View of the Expert–Lay

Knowledge Divide', in Lash, S. *et al.* (eds), *Risk, Environment & Modernity: Towards a New Ecology*, London: Sage, 1996

Young, I. M., *Justice and the Politics of Difference*, Princeton: Princeton University Press, 1990a

Young, I. M., *The Ideal of Community and the Politics of Difference*, in Nicholson, L. (ed.), *Feminism/Postmodernism*, London: Routledge, 1990b

Young, I. M., 'A Critique of Integration as the Remedy for Segregation', in Bell, D. and Haddour, A. (eds), *City Visions*, Harlow: Prentice Hall, 2000

Zamagni, S., 'Global Environmental Change, Rationality and Ethics', in Campiglio, L. *et al.* (eds), *The Environment after Rio*, London: Graham and Trotman, 1994

Zartaloudis, T., 'Without Negative Origins and Absolute Ends: A Jurisprudence of the Singular', 13 *Law and Critique* 2: 197–230, 2002

Zeleny, M. and Hufford, K., 'All Autopoietic Systems Must Be Social Systems (Living Implies Autopoietic. But Autopoietic does not Imply Living): An Application of Autopoietic Criteria in Systems Analysis', 14 *Journal of Social and Biological Structures* 3: 311–32, 1991

Zimmerman, M. E., 'Rethinking the Heidegger-Deep Ecology Relationship', *Environmental Ethics* 15 (Fall): 195–224, 1993

Index

For Product Safety Concerns and Information please contact our EU
representative GPSR@taylorandfrancis.com
Taylor & Francis Verlag GmbH, Kaufingerstraße 24, 80331 München, Germany

www.ingramcontent.com/pod-product-compliance
Lightning Source LLC
Chambersburg PA
CBHW050413280326
41932CB00013BA/1839

9 7 8 0 4 1 5 5 7 4 4 3 3